23 Update in Intensive Care and Emergency Medicine

Edited by J.-L. Vincent

Springer
Berlin
Heidelberg
New York
Barcelona
Budapest
Hong Kong
London
Milan
Paris
Santa Clara
Singapore
Tokyo

M. R. Pinsky J.-F. Dhainaut
A. Artigas (Eds.)

The Splanchnic Circulation

No Longer a Silent Partner

With 29 Figures and 16 Tables

Springer

Series Editor

Prof. Dr. Jean-Louis Vincent
Clinical Director, Department of Intensive Care
Erasme University Hospital
Route de Lennik 808, B-1070 Brussels, Belgium

Volume Editors

Prof. Dr. Michael R. Pinsky
University of Pittsburgh, Dept. of Anesthesiology
and Critical Care Medicine,
604 Scaife Hall, Pittsburgh, PA 15261, USA

Dr. Jean-Francois Dhainaut
Cochin Port-Royal University Hospital
Clinical Pharmacology and Critical Care Medicine
27 rue du Faubourg St. Jaques, F-75679 Paris, France

Dr. Antonio Artigas
Consorci Hospitalari Parc Tauli, Servei des Medicina Intensiva
Parc Tauli, E-08208 Sabadell, Spain

ISBN-13:978-3-642-79717-0 e-ISBN-13:978-3-642-79715-6
DOI: 10.1007/978-3-642-79715-6

Library of Congress Cataloging-in-Publication Data
The splanchnic circulation : no longer a silent partner / M. R. Pinsky, J.-F. Dhainaut, A. Artigas, eds. (Update in intensive care and emergency medicine ; 23) Based on the First International Symposium on Applied Physiology of the Peripheral Circulation, Sept. 23–24, 1995 in Barcelona.
Includes bibliographical references and index.ISBN-13:978-3-642-79717-0(hc:alk.paper) 1.Viscera-Blood-vessels-Pathophysiology Congresses.I. Pinsky, Michael R. II. Dhainaut, J.-F. (Jean-Francois) III. Artigas, Antonio. IV. International Symposium on Applied Physiology of the Peripheral Circulation (1st : 1995 : Barcelona, Spain) V. Series. [DNLM: 1. Splanchnic Circulation-physiology-congresses. 2. Gastrointestinal System-blood supply-congresses. 3. Critical Care-congresses. W 1 UP66H v. 23 1996] RC802.S67 1996 616.3'3-dc20 DNLM/DLC

Softcover reprint of the hardcover 1st edition 1995

SPIN: 10500565 19/3130-5 4 3 2 1 0 – Printed on acid-free paper

Preface

This monograph focuses on splanchnic function in health and disease. It represents a distillate of the communication that took place at the First International Symposium of Applied Physiology of the Peripheral Circulation, "Splanchnic Circulation: No Longer a Silent Partner." The individual chapters roughly follow the individual presentations and display in durable form the concepts and importance that this symposium achieved.

The concept for this annual symposium was the child of Antonio Artigas, who not only recruited sponsorship but also Jean-Francois Dhainaut and me to help with the organization and work. We chose the splanchnic circulation as the peripheral circulatory system to be presented first for many important reasons. Much new information has become available which demonstrates, as the title of the symposium implies, that splanchnic function has major influence on the overall expression of health and disease in humans. All aspects of splanchnic physiology, it seems, have been rediscovered to be dynamic, important, and complex in their interactions within individual tissues and among remote tissues and organs. It is hoped that after having reviewed this monograph the reader will agree that the splanchnic circulation and its organ systems are emerging as important aspects of critical illness and host-defense homeostasis.

The integration of diverse organ systems into a smoothly functioning organism that can adjust to life's stresses reflects the essence of homeostasis. Although one can marvel at the beauty of the system from afar, its processes and interactions among the organ systems reflect a complexity beyond this distant perspective. As part of the interaction among organs and organ systems, the microcirculation forms the pivotal interface for humoral processes. The complex nature of the vascular endothelial function is only now emerging. Endothelial cell function, both on the adlumenal side through its interaction with the vascular smooth muscle and on the lumenal side through its actions with formed circulating blood elements, plays a primary role in blood flow regulation and the expression of inflammation. Between the two perspectives, the whole organism and the microcirculation, lies integrative physiology.

In this monograph the integrative physiologic aspects of the splanchnic circulation in health and disease are addressed. The goal of this work is to summarize and focus recent insights into pathophysiologic processes into a single volume to be used both as a reference and as a guide for future clinical studies and basic scientific research. It is the product of the work of several authors from diverse backgrounds and with markedly different perspectives. Although a single idea was not developed consistently among chapters, the strength of this approach lies in insights being given from multiple perspectives and we feel that the goals which were set have been accomplished in a uniquely focused fashion.

Pittsburgh, September 1995 *Michael R. Pinsky*

Table of Contents

List of Contributors

A. G. N. Agustí
Servei Pneumología, Hospital Universitari Son Dureta, Universitat Illes Balears, Palma Mallorca, Spain

A. Artigas
Intensive Care and Internal Medicine Services, Hospital de Sabadell, Department of Molecular Biology and Physiology, Universitat Autonoma, Barcelona, Spain

F. Baigorri
Intensive Care and Internal Medicine Services, Hospital de Sabadell, Department of Molecular Biology and Physiology, Universitat Autonoma, Barcelona, Spain

R. Ballester
Servei Cardiologia, Hospital Universitari Vall Hebron, Departament de Fisiologia, Universitat Autonoma Barcelona, Barcelona, Spain

R. Bellomo
Intensive Care Unit, Austin Hospital, Heidelberg, Victoria 3084, Australia

G. Berlot
Istituto di Anestesia, Rianimazione e Terapia Antalgica, Cattedra di Terapia Intensiva, Università degli studi di Trieste, Trieste, Italy

R. A. Cabezas
Pulmonary and Critical Care Medicine Division, University of Texas Health Science Center, 6431 Fannin, Houston, TX 77030, USA

X. Calvet
Intensive Care and Internal Medicine Services, Hospital de Sabadell, Department of Molecular Biology and Physiology, Universitat Autonoma, Barcelona, Spain

D. De Backer
Departments of Intensive Care and Anesthesiology, Erasme University Hospital, Free University of Brussels, Route de Lennik 808, 1070-Brussels, Belgium

J.-F. Dhainaut
Medical ICU, Cochin Port-Royal University Hospital, 27 rue du Faubourg Saint Jacques, 75679 Paris Cedex 14, France

E. Domingo
Servei Cardiologia, Hospital Universitari Vall Hebron, Departament de Fisiologia, Universitat Autonoma Barcelona, Barcelona, Spain

M. R. Gilabert
Servei Cardiologia, Hospital Mutua Terrassa, Departament de Fisiologia, Universitat Autonoma Barcelona, Barcelona, Spain

A. Gullo
Istituto di Anestesia, Rianimazione e Terapia Antalgica, Cattedra di Terapia Intensiva, Università degli studi di Trieste, Trieste, Italy

A. Gurakar
Oklahoma Transplant Institute, Baptist Medical Center of Oklahoma, 3300 Northwest Expressway, Oklahoma City, OK 73112, USA

G. Gutierrez
Pulmonary and Critical Care Medicine Division, University of Texas Health Science Center, 6431 Fannin, Houston, TX 77030, USA

T. Hassanein
Loma Linda University Medical Center, Transplantation Institut, 11234 Anderson Street, Room 1405 Loma Linda, CA 92354, USA

J. Ibañez
Unitat Cures Intensives, Hospital Universitari Son Dureta, Universitat Illes Balears, Palma Mallorca, Spain

J. A. Kellum
University of Pittsburgh, Division of Critical Care Medicine, 200 Lothrop Street, Pittsburgh, PA 15213-2582, USA

K. Ley
University of Virginia Medical School, Department of Biomedical Engineering, Health Sciences Center Box 377, Charlottesville, VA 22908, USA

G. M. Matuschak
Division of Pulmonology, Department of Internal Medicine, St. Louis University School of Medicine, and, Department of Critical Care Medicine, St. John's Mercy Medical Center, St. Louis University Hospital, 3635 Vista Ave. at Grand Blvd., St. Louis, MO 63110-0250, USA

A. Meier-Hellmann
Department of Anesthesiology, Friedrich-Schiller University Hospital, Bachstraße 18, 07740 Jena, Germany

A. Mignon
Medical ICU, Cochin Port-Royal University Hospital, 27 rue du Faubourg Saint Jacques, 75679 Paris Cedex 14, France

M. R. Pinsky
Department of Anesthesiology and Critical Care Medicine, University of Pittsburgh, Pittsburgh, PA 15261, USA

K. Reinhart
Department of Anesthesiology, Friedrich-Schiller University Hospital, Bachstraße 18, 07740 Jena, Germany

R. Rodriguez-Roisin
Servei de Pneumologia i Allèrgia Respiratòria, Hospital Clínic, Villarroel 170, 08036 Barcelona, Spain

J. A. Russell
Department of Medicine, St. Paul's Hospital, Associate Professor of Medicine, Division of Critical Care Medicine, University of British Columbia, Vancouver, B.C., Canada

H. K. F. van Saene
Department of Intensive Care, Academic Medical Center, Amsterdam, and, Department of Medical Microbiology, University of Liverpool, P.O. Box 147, Liverpool L69 3BX, UK

D. Schmartz
Departments of Intensive Care and Anesthesiology, Erasme University Hospital, Free University of Brussels, Route de Lennik 808, 1070-Brussels, Belgium

C. P. Stoutenbeek
Department of Intensive Care, Academic Medical Center, Amsterdam, and, Department of Medical Microbiology, University of Liverpool, P.O. Box 147, Liverpool L69 3BX, UK

D. Thébert
Medical ICU, Cochin Port-Royal University Hospital, 27 rue du Faubourg Saint Jacques, 75679 Paris Cedex 14, France

B. Togores
Servei Pneumología, Hospital
Universitari Son Dureta,
Universitat Illes Balears,
Palma Mallorca, Spain

D. H. Van Thiel
Oklahoma Transplant Institute,
Baptist Medical Center of
Oklahoma, 3300 Northwest
Expressway, Oklahoma City,
OK 73112, USA

J.-L. Vincent
Departments of Intensive Care and
Anesthesiology, Erasme University
Hospital, Free University of
Brussels, Route de Lennik 808,
1070-Brussels, Belgium

Regional Blood Flow Distribution

M. R. Pinsky

Introduction

Blood flow to the tissues of the body is regulated by complex and redundant homeostatic systems that interact to ensure adequate delivery of oxygen (O_2) to the metabolically active tissues. Since metabolic requirements are rarely constant throughout the day, blood flow also varies. In fact, there is a remarkable coupling of cardiac output to global O_2 consumption (VO_2), such that changes in VO_2 induce proportionally similar changes in cardiac output (Fig. 1a) in a fashion completely analogous to changes in minute ventilation following changes in CO_2 production. The cause of this global coupling of cardiac output to VO_2 is the peripheral circulatory changes among organs and vascular beds to match regional blood flow with regional metabolic demand (Fig. 1b). The mechanisms by which this occurs are only partially known. Furthermore, it is their uncoupling which signifies one of the most common clinical signs of sepsis in the critically ill. Possibly through better understanding these different mechanisms and how disease may alter them we can come to a better understanding of the pathophysiological mechanisms which underlie severe illness and develop novel strategies to correct them.

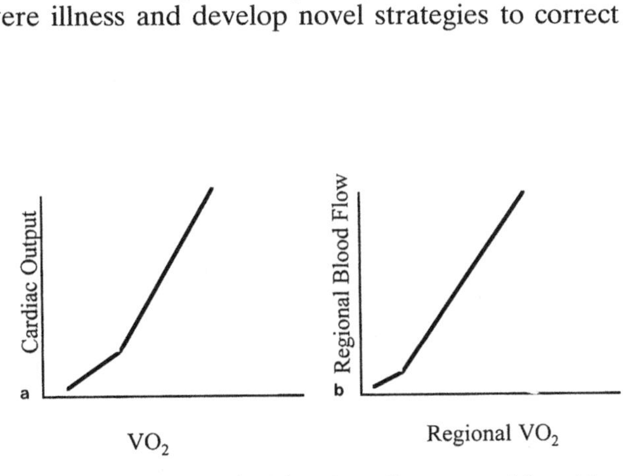

Fig. 1. Relationship between global oxygen uptake (*VO_2*) and cardiac output (**a**) and between regional metabolic demand of a given tissue bed and its total blood flow (**b**). The reason that cardiac output tracks VO_2 is that regional blood flow tracks regional metabolic demand

Mechanisms Regulating Local Blood Flow on a Macrovascular Level

Blood flow to specific vascular beds and organs is regulated by three different but interrelated processes. These processes are loosely grouped as local autoregulatory control, global autonomic control, and passive local pressure-flow relationships. Local autoregulatory processes include local release and action of vasodilating substances, capillary recruitment, and local autonomic feedback control.

As tissues become relatively ischemic, meaning that the supply of O_2 is inadequate to meet the ongoing metabolic demands, reducing substances necessary for production of adenosine triphosphate (ATP) from adenosine diphosphate (ADP) and inorganic phosphate (P_i) become the rate-limiting substances. Cells can transiently bypass this deficiency by combining two ADP molecules to form one ATP and adenosine monophosphate (AMP). If O_2 flux is still inadequate, ADP combines with AMP to form ATP and adenosine. Adenosine is lipid permeable and freely diffuses out of the cytosol and is lost. It is also one of the most potent smooth muscle relaxing substances known. Thus, by releasing adenosine from ischemic cells, reflex vasodilatation can occur without the need for intact autonomic circuits or systemic endocrinological responses [1]. Adenosine is very rapidly broken down in the circulation by the enzyme adenosine deaminase. Apparently adenosine deaminase is in the circulation for just this reason, to prevent the systemic vasodilating effects of adenosine from occurring following its local release.

However, adenosine-induced vasodilatation is for several reasons neither the only mechanism responsible for increased organ blood flow nor the most common. First, loss of adenosine from the cell represents an inefficient mechanism to regulate local blood flow because it is rapidly degraded and is expensive in the metabolic sense of the word to synthesize de novo. Second, most increased metabolic demand follows a programmed scenario, common to many stresses. For example, increased muscular activity with walking or running, increased gut metabolism with digestion, and increased cerebral metabolism with mentation. Local autonomic control exists for just these circumstances. It induces vasodilatation of resistance vessels in regions assumed to be needing of increased blood flow. Since this local autonomic autoregulation is more generalized than adenosine release, the increase in blood flow often exceeds the region of tissue with increased metabolic activity. One manifestation of increased autonomic tone is an increase in the number and diameter of perfused capillaries in the tissue in question. These processes represent the final microvascular result of the increased regional blood flow.

Finally, as organ perfusion pressure increases, organ blood flow also increases. The gain of this system, that is, the degree to which a given increase in arterial pressure increases organ blood flow differs among organs [2]. Thus, changing mean arterial pressure induces blood flow redistribution

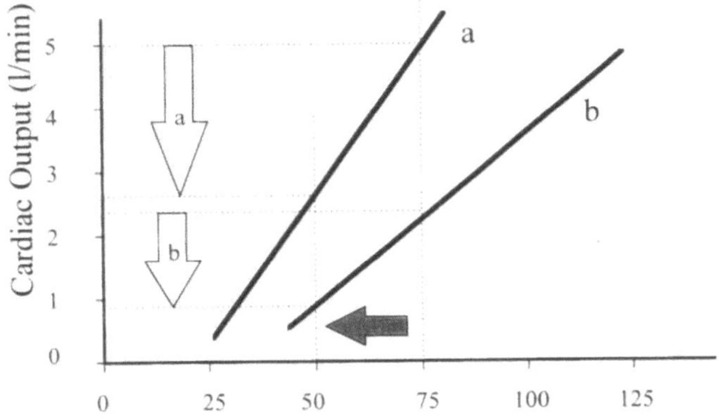

Fig. 2. Perfusion pressure versus organ blood flow of two vascular beds (*a*, *b*) with differing resistances and zero-flow intercepts. Note that decreasing organ perfusion pressure from 75 to 50 mmHg decreases flow to vascular bed *a* more than to vascular bed *b* because of these differences in passive pressure-flow characteristics between the two vascular beds

among organs due to differences in their passive pressure-flow relationships. These points are illustrated in Fig. 2 for two hypothetical parallel vascular beds with differing pressure-flow relationships. As perfusion pressure decreases, the flow reduction to the respective beds differs both in magnitude and in relative proportion. Unfortunately, both stress and sepsis can alter the passive pressure-flow relationships among organs in a nonuniform fashion. Thus, changes in organ perfusion pressure may have one effect on blood flow redistribution during normal conditions and an entirely different effect in patients with, for example, trauma or sepsis.

Global autonomic tone alters the entire circulation in very specific ways. In essence, the body behaves as if it already knew what the probable problems were in the body and induces a nonspecific stress response which decreases blood flow to skin, gut, kidney, and noncontracting muscle while preserving blood flow to brain, heart, diaphragm, and other contracting muscles. This centrally mediated blood flow redistribution has been elegantly described using microsphere techniques which allow one to define blood flow distribution among tissues as discrete points in time as the whole body is stressed in various ways [3]. When this control is lost, the cardiovascular system is less able to adapt to stress without allowing some tissue beds from becoming hypoxic. Abnormal O_2 metabolism for whatever reason is referred to as dysoxia in this paper. For example, Cain et al. [4] demonstrated that blocking α-adrenergic tone resulted in a higher global oxygen delivery (DO_2) when the dysoxic threshold was reached by hypoxic hypoxia in dogs than if α-adrenergic tone was left intact. Numerous studies since then have shown roughly similar results. Reversing this model, Maginniss et al. [5] demonstrated with augmented stress response prior to reducing DO_2 that O_2

extraction was higher for any level of DO_2 than if no prior stress was given. Thus, animals so stressed were able to maintain adequate levels of tissue oxygenation despite lower absolute levels of DO_2 than nonstressed animals. Presumably the heightened sympathetic tone maintained arteriolar blood flow to metabolically active tissues while reducing blood flow to nonactive ones.

Mechanisms Regulating Local Blood Flow on a Microvascular Level

However, even if arteriolar blood flow does go to the correct tissues in proportion to their needs, there is no guarantee that the downstream capillaries receive this blood in an amount proportional to the needs of their tissues. This statement may seem odd but reflects the realization that capillary blood flow does not always follow arterial blood inflow. Specifically, capillary blood flow is not always what one would expect based on simple principles of capillary cross-sectional area, capillary lumenal volume, capillary density, and tissue-capillary ratios as described using Krough tissue cylinder model. The reasons for this discrepancy are multiple, and all are important for understanding regional blood flow in critically ill patients. The first and easiest reason for this discrepancy relates to the fact that in the best of conditions capillary hematocrit is approximately 30% less than arterial hematocrit. Since an obligatory plasma layer must interface between the blood stream and the vascular endothelium [6], relative capillary dilution must occur. This phenomena is referred to as the Fahreaus effect [6]. Clearly, similar plasma layers exist at the interface of all blood stream-endothelial junctions – it is just that in the capillary the cross-sectional diameter is so small that this film of plasma becomes quantitatively significant. Second, precapillary sphincters gate red cell access to capillaries, decreasing capillary hematocrit even further. As precapillary sphincter tone increases, capillary hematocrit decreases until flow to the capillary ceases all together. Complicating this process of active gating of red cells even further, streaming effects of red cells limit red cell egress into capillaries that take off in right angles to the direction of flow of the main stream of arterial blood. Certain end-capillaries, such as those in the gut mucosa take off at right angles to the stream of blood flow, whereas muscular tissues tend to have capillary beds as end-on capillaries.

It is not entirely clear, however, the extent to which these differences in capillary hematocrit alters tissue O_2 delivery. One can easily argue that since the flow of fluid through the center of the capillary exceeds the rate of flow of fluid adjacent to the endothelium, actual red cell delivery to the tissues may exceed "capillary hematocrit." In this context we have referred to capillary hematocrit as that calculated from inspection of the capillary and whole blood hematocrit as that measured when blood is collected from veins. Capillary hematocrit can be calculated from the ratio of the total volume of fluid in the capillary and the volume of red cells in the capillary. When measured using microvascular imaging techniques, capillary hematocrit may be as low as 15% despite a whole body hematocrit of 45% [7, 8].

As metabolic activity increases in a tissue, its microvascular bed changes its flow characteristics, such that capillary hematocrit increases until it approaches a level predicted by the Fahreaus effect. One may argue that this capillary hematocrit is less important than might be expected because the total flux of red cells through the capillaries "behaves" as if it had a whole blood hematocrit. This means that blood collected from capillaries approximates whole blood hematocrit. However, within the capillary the intercorpuscular distance increases as the capillary hematocrit decreases. This red cell free space may quickly become hypoxemic since it has no O_2 reservoir of hemoglobin to sustain its pO_2. In metabolically active tissues this hypoxic zone between red cells results in impaired O_2 uptake by the tissues and decreased oxidative phosphorylation. The tissues behave as if the capillary pO_2 were less than it actually is, and in fact end-capillary pO_2 when measured in venous blood is greater than tissue pO_2 for the same vascular bed. Increases in capillary hematocrit increase O_2 availability and are reflected in microvascular studies as both recruitment and distention of capillaries. Furthermore, in anemic states precapillary sphincter tone decreases, tending to maintain capillary hematocrit constant despite changing whole blood hematocrit values. Thus, varying capillary hematocrit despite constant whole blood hematocrit or vice versa probably reflects both normal adaptive mechanisms and potential pathophysiological mechanisms to prevent DO_2 from becoming VO_2 despite adequate levels of local tissue DO_2.

The implication of this microvascular modulation of DO_2 is that actual DO_2 may vary quite closely with local variations in VO_2 and over a wide range of values despite minimal changes in global O_2 transport. In fact, these microvascular alterations in capillary hematocrit may well explain many of the normal variations in O_2 extraction seen among different tissues. Furthermore, since decreasing whole body hematocrit increases capillary hematocrit by reducing the capillary gating effect on red cell flow, some of the observed increase in O_2 extraction seen with anemia may actually reflect greater DO_2 on a microvascular level (increased capillary hematocrit without tissue ischemia) than increased O_2 gradient-induced increased O_2 extraction (obligatory tissue ischemic zones) [9].

Basic Questions about the Peripheral Circulation

Assuming that microcirculatory flow may not follow macrocirculatory flow, and that passive pressure-flow relationships among different vascular beds not only differ but may change in a differential fashion in response to stress and disease, one can ask several nihilistic questions. I attempt to answer these briefly below.

1. Do global measures of blood flow predict regional measures of blood flow? This means that if one measures a cardiac index and O_2 carrying

capacity that is in the normal range, can one assume that regional blood flow is also adequate? Clearly, the answer to this question is no. Maldistribution of flow among and within organs may mask circulatory insufficiency in sepsis and other states.
2. Does maintenance of regional blood flow prevent organ-system dysfunction? If tissue hypoxia were the only cause of organ-system dysfunction, maintenance of regional blood flow would go a long way in ensuring tissue viability. To the extent that cytotoxic substances induce organ-system dysfunction it is not all together clear that a normal or even supranormal regional blood flow would ensure preservation of organ-system function.
3. Does restoration of regional blood flow reverse organ-system dysfunction? This is the central tenant for resuscitation and resuscitative maneuvers. Regrettably, little evidence supports this hypothesis, especially in patients with sepsis. Clearly, hypoperfusion shock states due to decreased cardiac output (hypovolemia, cardiac pump failure, obstructive processes) respond well to restoration of blood flow with return of organ-system function of minutes (brain), to hours (heart), to days (kidney). However, in sepsis and other systemic inflammatory response states all studies to date document a relative resistance of organ dysfunction to respond favorably to resuscitation maneuvers.
4. Are there any markers of adequacy of regional blood flow? To a large extent the answer depends on how one defines adequacy. Clearly, mentation documents adequate cerebral blood flow, a urine output above some minimal level reflects adequate glomerular blood flow, and normal gut peristalsis and ventricular pump function reflect adequate gut mesenteric and coronary blood flow. Still, most critically ill patients manifest outward signs of organ-system dysfunction which may not reflect ischemic dysfunction nor improve markedly if due to ischemic dysfunction during the early phases of resuscitation despite appropriate therapy. At present numerous markers of the adequacy of regional blood flow are being explored. For the gut gastric tonometry is now popular but awaits further studies before its widespread use can be recommended. Clearly, more basic science work both in the laboratory and the clinic needs to be carried out prior to our developing a process-specific monitoring and treatment strategy.

Models To Assess Regional Blood Flow

Regional blood flow can be measured either directly or indirectly using a variety of techniques. Each technique has its strengths and weaknesses. These issues are discussed briefly below because they impact on the relevance of data presented both here and elsewhere in this monograph.

Direct Measures

Numerous techniques have been developed over the past 30 years to measure continuously and on-line blood flow and flow velocity in both isolated and intact animals and man. Flow probes measuring either induced electric fields (electromagnetic) or Doppler shift (ultrasonic) have been developed that are accurate and can be used in chronic models and acutely to measure blood flow continuously and over wide ranges of flow. Their primary strengths are their accuracy and the ability to sample flow data continuously, thus allowing one to assess dynamic changes in flow as pressure or resistance is rapidly varied. These techniques are limited, however, in several important ways. First and foremost, they are invasive in nature, requiring exposure of the vessels to be studied. Furthermore, they measure only one vascular bed and then only the macrovascular flow. If one wishes to measure blood flow redistribution, it is necessary to insert several flow probes on different vessels. The practical limits of flow measurement are for vessels 1 mm in diameter or larger, with the accuracy of measurement increasing as vessel size and flow velocity increase.

To circumvent the limitation of single vessel blood flow studies microsphere techniques have been developed in which an emulsion of microspheres is injected into the left atrium and allowed to embolize throughout the body. The microspheres are sized to impact in the smallest arterioles. A paired sample of blood is simultaneously collected from the arterial tree at a known rate to serve as a reference "mock" organ. Knowing the total number of microspheres injected and the number collected in the mock organ, one can easily calculate the proportion and total blood flow in any tissue sample. The microspheres are labeled with either a radioactive marker or a color. Microsphere analysis of blood flow distribution allows one to examine both within and among organ blood flow distribution. As with flow probe technologies, however, this technique is also limited in its application. One obtains only a "snapshot" of the circulation at one time. Although several "snapshot" injections of microspheres can be given, the total number is limited to between four and and may become inaccurate at the high end because of prior embolization-induced blood flow redistribution.

Indirect Measures

Noninvasive and relatively noninvasive markers of tissue metabolism have also evolved which allow the assessment of regional blood based on the assumption that adequate regional blood flow should manifest itself as good tissue oxygenation in a nonacidotic environment. In this regard increased interest has developed for the use of the multiple electrode oxygen sensor technique originally described by Lund. This apparatus uses an array of multiple micro-O_2 sensors configured on a compressed sensing head. This allows one to measure local surface pO_2 of multiple adjacent regions of tissue both a one point in time and then over time. Such studies have yielded significant insights into the process of tissue oxygenation. For example, tissue pO_2 is not one value but a range of values that distribute themselves around a normalized mean. Furthermore, the specific value for given loci varies about this mean value and is not constant. Finally, specific loci have specific pO_2 states in which they exist, such that they vary among these given states in a steplike fashion (digital) rather than varying in a gradual one (analogue). These findings alone have led researchers to conclude that tissue capillary flow varies continually in specific steps as metabolic demand varies, such that there is no specific tissue pO_2 but rather a distribution and range of tissue pO_2 values. The surface multiple electrode technique is limited, however, in assessing only surface pO_2. Thus, tissues deep to organ surfaces, such as contracting muscle, renal parenchyma, and brain may not be assessed accurately by this technique. Still, as an adjutant technique to those for direct measurement it provides valuable insights into tissue wellness and the pathophysiology of tissue dysfunction in critical illness.

Numerous techniques assessing acid-base status have been studied as indirect measures of regional blood flow. The thinking here is that tissue acidosis reflecting either impaired CO_2 removal by a low-flow system or inadequate oxidative phosphorylation inducing anaerobic metabolism induce elevations in tissue pCO_2 and increased lactic acid production. Furthermore, these changes are presumed to reflect tissue dysoxia, and their correction, it is hoped, would reverse organ-system dysfunction. Examples of these techniques include gastric tonometry, in which pCO_2 of the gastric fluid is assumed to be in equilibrium with interstitial fluid, allowing calculation of interstitial pH (pHi) by the Henderson-Hasselblach equation, assuming tissue bicarbonate equals blood bicarbonate. One need not make these assumptions, however. Direct measurements of tissue pH, pO_2, and pCO_2 can be made using microelectrodes. Although invasive and perhaps inducing tissue injury, these techniques benefit from lack of assumptions in the measurement of pHi. Whether measures of tissue pH or other acid-base markers will be of clinical relevant is still being disputed. Data to date suggest that sustained tissue acidosis is a grave prognostic sign. However, studies aimed at improving circulatory status with an eye toward restoration of normal acid base status have not demonstrated beneficial effects in critically ill patients.

Control of Regional Blood Flow Distribution

Increased autonomic tone in response to stress alters vasomotor tone in an attempt to optimize blood flow distribution to metabolically active tissues necessary for immediate survival. It does this by both altering arterial inflow resistance and tone in the venous capacitance vessels. This coordinated process results in a different circulatory response depending on whether blood volume is reduced while stress is maintained constant, or blood volume is constant but metabolic demand is increased. The former example represents progressive low-output shock as may occur during hypovolemia, cardiac failure, or circulatory obstruction, whereas the latter represents the condition of increased metabolic demand in the setting of impaired circulatory reserve.

In an anesthetized operated canine model Schlichtig et al. from our laboratory measured blood flow to gut, kidney, liver, and whole body simultaneously using flow probes as circulating blood volume was progressively reduced by hemorrhage [10]. This study demonstrated that blood flow decreased differently from gut, liver, kidney, and the remainder of the body. Fractional blood flow to the liver decreased the most, while renal and mesenteric blood flow decreased less, but still more than the remainder of the body. Furthermore, different vascular beds have differing levels of O_2 extraction efficiency. The kidney extracts O_2 poorly, the liver very well, and the gut intermediate between the two. Based on these observations we concluded that blood flow tends to be distributed away from organs with better O_2 extraction capacities, allowing global DO_2 to decrease to the lowest possible levels before dysoxia ensues. Interestingly, in a companion study specifically examining renal O_2 uptake we observed a highly linear relationship between renal blood flow and renal VO_2 [11]. These data are understandable when one considers that renal metabolic work is proportional to the filtered sodium load. Since the primary determinant of filtered sodium load is renal blood flow, it follows that not only is blood flow redistributed away from the kidneys toward the rest of the body during progressive hemorrhage, but that intrinsic renal metabolic demand also decreases, thus minimizing the chances of renal hypoperfusion inducing acute renal failure. This process is limited, however, once renal blood flow falls low enough because the kidney then also becomes ischemic and starts to excrete rather than consume lactate.

As an extension of these studies Kramer et al. from our laboratory examined the pressure-flow characteristics of renal, hepatic, and mesenteric arteries of an intact canine model in response to abrupt decreases in mean arterial pressure induced by transient occlusion of the inferior vena cava [12, 13]. They demonstrated that different vascular beds have different slopes of their pressure-flow relationships and different zero-flow pressure intercepts. The kidney had the lowest resistance, whereas the liver had the highest. Thus, similar increases in mean arterial pressure induce differing degrees of increased blood flow in various vascular beds. The kidney shows the greatest increase in blood flow and the liver the least. Accordingly, some of the observed blood flow redistribution seen in hypovolemic shock if associated with hypotension may reflect passive vascular pressure-flow relationship differences among organs rather than active changes in vasomotor tone.

The primacy of perfusion pressure in determining organ blood flow was underscored recently by a canine study by Romand et al. from our laboratory. In this model moderate hemorrhage was induced to place the circulation in a stressed but compensated state. Then a remote increase in metabolic demand was induced by electrical stimulation of the femoral nerves [14]. In this model gut, kidney, and lower extremity blood flow decreased during progressive hemorrhage as prior studies had demonstrated, but blood flow to neither the gut nor kidney decreased further during mock exercise unless mean arterial pressure also decreased. During stimulation periods when mean arterial pressure remained constant cardiac output and lower extremity blood flow increased, lactic acidosis ensued, but no reduction in gut or kidney blood flow occurred. However, in the same animals if stimulation increased a hypotensive response, the both gut and kidney blood flow also fell and did so in parallel fashion in phase with the falling mean arterial pressure. Based on these observations we concluded that blood flow redistribution occurs early during stress states but is limited in its magnitude. Once established, further reductions in gut and renal blood flow occur only if arterial pressure also decreases.

Regional Blood Flow Distribution During Sepsis

In sepsis complex inflammatory processes ensue which profoundly alter the homeostatic circulatory process of the body. The proposed mechanisms operative in this response and some of the putative mediators are the following:

- Decreased adrenergic receptor sensitivity
  - Internalization of adrenergic receptors
  - Reduced G protein activity
- Increased microvascular resistance
  - Microemboli of formed cellular elements
  - Endothelial swelling
  - Release of vasoconstricting substances: histamine, thromboxane A_2, prostaglandin $F_{2\alpha}$, platelet-activating factor, leukotrienes
- Decreased vasomotor tone
  - Induceable nitric oxide syntase upregulation
  - Release of vasodilating substances: prostacyclin, leukotrienes

Clinically there appears to be a dissociation between regional metabolic demand and organ blood flow and between nutritive blood flow and cellular uptake and utilization of O_2. The generalized inflammatory response induces dysfunction throughout the body and has been modeled as a panendothelial disease. We have referred to it as malignant intravascular inflammation to underscore the mechanistic nature of the syndrome [15]. Sepsis alters vasomotor tone and the distribution of blood flow among and within organs. The acute response to an overwhelming stress such as bacteremia, trauma, pancreatitis, and anaphylaxis is for the adrenergic receptors on the endothelial cell surfaces to internalize markedly reducing the adrenergic responsiveness of the circulation. Activated endothelial cells and formed cellular elements in circulation express vasoconstricting lipid mediators, such as thromboxane, prostaglandin $F_{2\alpha}$, platelet-activating factor, and leukotrienes. These mediators have been hypothesized to induce pathological vasoconstriction although the data supporting these conclusions are mainly circumstantial. Similarly, vasodilating substances are also released including nitric oxide and certain leukotrienes. Here, however, the data are more convincing that these

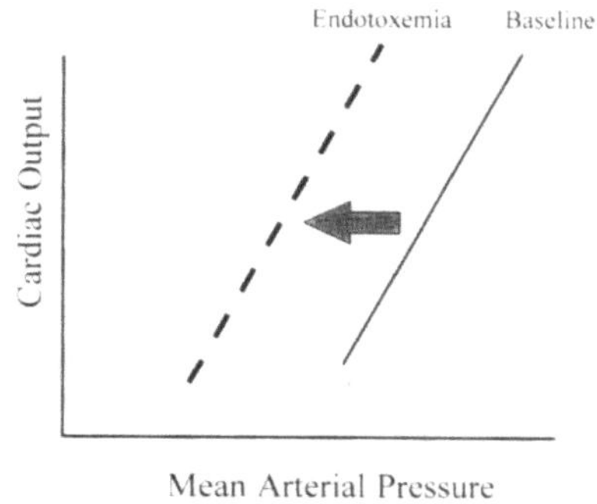

Fig. 3. The relationship between mean arterial pressure and cardiac output as cardiac output is varied by rapidly changing intravascular volume under baseline (*closed circles*) and endotoxemic conditions (*dotted line*). Note that the slope of the pressure-flow relationship is not altered (no change in resistance), but that the zero-flow pressure intercept decreases during endotoxemia, consistent with a loss of precapillary sphincter tone. (Modified from [16])

mediators induce the vasodilated state commonly seen in resuscitated septic patients. The physiological effects of this cacophony of disregulated vasoactive mediator release is the septic state with its profound alterations in regional blood flow regulation.

When examining the peripheral vascular determinants of the hemodynamic response to acute endotoxemia in a canine model, Pinsky and Matuschak demonstrated that the arterial pressure-flow relationship for the whole body was remarkably preserved, but that the zero flow pressure (critical closing pressure of the circulation) was markedly reduced [16] (Fig. 3). If one assumes that the slope of the pressure-flow relationship is determined by small artery and arteriolar tone whereas the zero-flow pressure is determined by precapillary sphincter tone, these data suggest that loss of precapillary sphincter tone not arterial vasomotor tone is the primary process inducing hypotension in sepsis. Pinsky eventually demonstrated that the gross peripheral vascular response to acute endotoxemia can be abolished by pretreatment with ibuprofen, a cyclo-oxygenase inhibitor, but that the metabolic manifestations of acute endotoxemia (acidosis, increased VO_2, and lactate production) are unaltered [17]. Furthermore, Kramer et al. from our laboratory demonstrated in their study cited above that when animals are given endotoxin the pressure-flow relationships among the various vascular beds tend to approach each other. Since some had greater tone prior to endotoxemia, whereas others had less tone, this shift appeared to cause some vascular beds to increase their tone and others to decrease it. The data do not allow us to conclude that tone is actively increased in some beds since an equally reasonable hypothesis is that active vasodilation is blocked by the inflammatory process. Interestingly, Bellomo et al. from our laboratory recently demonstrated that the addition of moderate exogenous adrenergic stimulation (norepinepherine) does not alter the pressure-flow relationships induced by acute endotoxemia in this model (unpublished data).

Resuscitation and Regional Blood Flow in Sepsis

Does resuscitation with the aim of restoring regional blood flow to normal levels restore organ function in sepsis? This is at present an unanswered question, although several investigators have given us clues into the right answer. Fink et al. demonstrated that the increase of cardiac output to supranormal levels to restore mesenteric blood flow to baseline levels in endotoxin dogsrestores gut pHi and intestinal permeability to normal levels [18]. Underresuscitated animals in this study demonstrated marked decreases in gut pHi decreased and increases in gut permeability. However, important and different findings were reported by Vallet et al. when they examined other markers of tissue function [19]. Using the multiple electrode oxygen sensor probe placed on the gut mucosa and serosa as well as mesenteric flow probes, they demonstrated that the increase in mesenteric flow induced by resuscitation in endotoxic animals goes to the serosal surface and not the mucosal surface. In fact, mucosal ischemia and acidosis developed in their model despite very high mesenteric blood flows. Thus, it is not clear whether resuscitation even to levels of blood flow which restore gut blood flow to normal would either represent or reverse gut mucosal ischemia. Since gut mucosal ischemia is considered a major contributing factor in maintaining the inflammatory state, this issue is relevant to clinical practice.

Conclusion: In Search of Measures of Tissue Dysoxia

In a previous editorial I examined the tenants which underscore our use of global O_2 supply-demand analysis. I came to the conclusion, based on data not dissimilar from those presented here, that global measures of either DO_2 or VO_2 do not reflect regional organ-system function, nor can they be used for trend monitoring [20]. Regrettably, we still do not have reliable markers of regional metabolic function, and no marker has been shown to be superior in prognostication to simple measures of global hemodynamic status. Future measures of tissue wellness need to focus on regional function from a metabolic perspective. The extent to which such markers exist and the extent to which they are useful in clinical decision making are two separate matters. However, markers of regional function are essential if we are to assess regional blood flow and regional metabolic function with an eye toward titration of therapy. Still, we have yet to prove the primary question that makes us interested in regional blood flow in the first place: is regional blood flow the primary determinant of organ function and survival in the resuscitated patient? Until we can answer this question, there is little new that resuscitative efforts can accomplish which will be of proven clinical value.

References

1. Berdenheur H, Schrader J (1986) Supply-to-dmand rtio for oygen determines formation of adenosine by the heart. Am J Physiol 250:H162–H173
2. Kramer DJ, Stein KL, Schlichtig RA, Armendariz E, Lanier A, Pinsky MR (1989) Pressure flow relationships in the superior mesenteric and hepatic arteries in endotoxic shock. Chest 96:2933
3. Magder S, Erian R, Roussos C (1986) Respiratory muscle blood flow in oleic acid-induced pulmonary edema. J Appl Physiol 60:1849–1856
4. Cain S (1978) Effects of time and vasoconstrictor tone on O_2 extraction during hypoxic hypoxia. J Appl Phsyiol 45:219–224
5. Maginniss LA, Connolly H, Samsel RW, Schumacher PT (1994) Adrenergic vasoconstriction augments tissue or extraction during reductions in O_2 delivery. J Appl Physiol 76:1454–1461
6. Klitzman B, Duling B (1979) Microvascular hematocrit and red cell flow in resting and contracting striated muscle. Am J Physiol 237:H481
7. Cokelet GR (1992) Speculation on a cause of lower vessel hematocrits in the microcirculation. Microcirculation 2:1–18
8. Desjardins, Duling B (1990) Heparinase treatment suggests a role for the endothelial cell glycocalyx in regulation of capillary hematocrit. Am J Physiol 258:H747
9. Dantzker D, Gutierrez G, Foresman B (1991) Oxygen supply and utilization relationships: a reevaluation. Am Rev Respir Dis 143:675–679
10. Schlichtig RA, Kramer DJ, Pinsky MR (1991) Flow redistribution during progressive hemorrhage is a determinant of critical O_2 delivery. J Appl Physiol 70(1):169–178
11. Schlichtig RA, Kramer DJ, Boston JR, Pinsky MR (1992) Renal O_2 consumption is supply-dependent at all levels of renal O_2 delivery during progressive hemorrhage. J Appl Physiol 70(1):1957–1962
12. Kramer DJ, Pinsky MR, Schlichtig RA (1989) The relation between regional and systemic oxygen delivery (DO_2) during progressive hemorrhage. Am Rev Respir Dis 139:A540
13. Schlichtig RA, Kramer DJ, Pinsky MR (1989) Regional and systemic oxygen delivery and consumption during progressive hemorrhage. Chest 94:755
14. Romand J, Attewell J, Pinsky MR (1995) Increases in peripheral O_2 demand affect blood flow distribution in hemorrhaged dogs. Am J Respir Crit Care Med (in press)
15. Matuschak GM, Pinsky MR (1989) Effect of positive-pressure ventilation frequency on hepatic blood flow and performance. J Crit Care 4(3):153–165
16. Pinsky MR, Matuschak GM (1986) Cardiovascular determinants of the hemodynamic response to acute endotoxemia in the dog. J Crit Care 1(1):18–31
17. Pinsky MR (1992) The cardiovascular response in canine endotoxic shock: effect of ibuprofen pretreatment. Circ Shock 37:323–332
18. Fink M, Kaups KL, Wang H, Rothschild HR (1991) Maintenance of superior mesenteric arterial perfusion prevents increased intestinal mucosal permeability in endotoxic pigs. Surgery 110:154–161
19. Vallet B, Lund N, Curtis SE, Kelly S, Cain SM (1994) Gut and muscle tissue PO_2 in endotoxemic dogs during shock and resuscitation. J Appl Physiol 76:793–800
20. Pinsky MR (1994) Beyond global O_2 supply-demand relations: in search of measures of dysoxia (editorial). Intensive Care Med 20:1–3

Gut Metabolism and Anion Flux

J.A. Kellum and R. Bellomo

Introduction

In the traditional view of metabolic acid-base derangements homeostasis involves mostly the work of the kidneys and the lungs. Recent conceptual developments, however, point to the liver as a major organ involved in pH homeostasis [1]. Similarly, the conventional view of the gut as a source of acidosis has undergone considerable challenge in certain settings (such as sepsis). This chapter explores these new developments in a basic science as well as a clinical context. The information contained in this chapter is intended for both the critical care researcher and practitioner.

Metabolic Acidosis in Sepsis

Conventional wisdom has it that lactic acid is the primary source of excess metabolic hydrogen ions occurring in sepsis [2]. In this view lactic acid is released primarily from the muscle and the gut as a consequence of tissue hypoxia. Moreover, the amount of lactate produced is felt to correlate with "the total oxygen debt, the magnitude of the hypoperfusion, and the severity of shock" [3]. In this conventional view the primary anatomic sources of lactate are thought to be the muscle and the gut. The assumption that the muscle is involved is based on a long tradition of looking to the muscle as a source of lactic acid during exercise. In this setting the muscle may be the only significant source of lactate. This observation may be safely generalized to certain patients with trauma or seizures, but here the generalization should end. During sepsis, even with profound shock, resting muscle does not produce lactate. In fact studies carried out by Van Lambalgen and coworkers [4] and more recently by Cain and Curtis [5] show that the muscle may actually consume lactate during endotoxemia.

Data concerning the gut are less clear. There is little question that profoundly ischemic gut can release lactate. However, does the gut release lactate during sepsis if gut perfusion is maintained? There is evidence, in fact, to support the notion that it does just the opposite. These studies have shown that the gut is neutral to, or even takes up, lactate in such conditions [4, 6]. Perfusion is likely a major determinant of gut lactate metabolism. In a canine

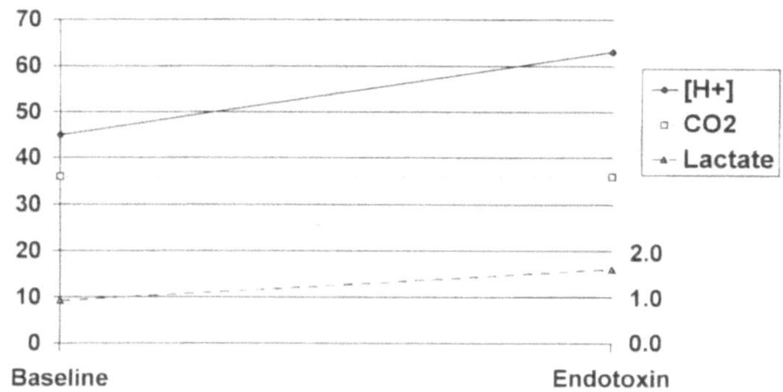

Fig. 1. Changes in $[H^+]$ (*solid line*; nmol/l), pCO_2, (*dotted line*; mmHg), and lactate (*dashed line*; mmol/l) during early endotoxemia (30 min). Data obtained during the unresuscitated state

model of sepsis using endotoxin gut lactate production could not be shown when flow was maintained with dopexamine hydrochloride [5].

What then is the source of lactate in sepsis? In our canine model of experimental sepsis using endotoxin infusion we saw that in the acute endotoxic condition arterial lactate increased, as did arterial hydrogen ion concentration, although the change in hydrogen ion concentration exceeded that attributable to lactate (Fig. 1). When we examined the flux of lactate across various organs using ultrasonic flow probes to quantify flow accurately, indeed the gut and muscle did not contribute to lactic acidosis (Fig. 2). In fact, both muscle and gut tended to take up lactate in this condition. Surprisingly, the lung emerged as the major source of lactate [6]. Production of lactate by the lung is not a new concept. Other groups have reported lung lactate production since Sayeed's work [7] in the early 1980s [8]. More recently Gutierrez and coworkers reported a transpulmonary gradient in patients with sepsis

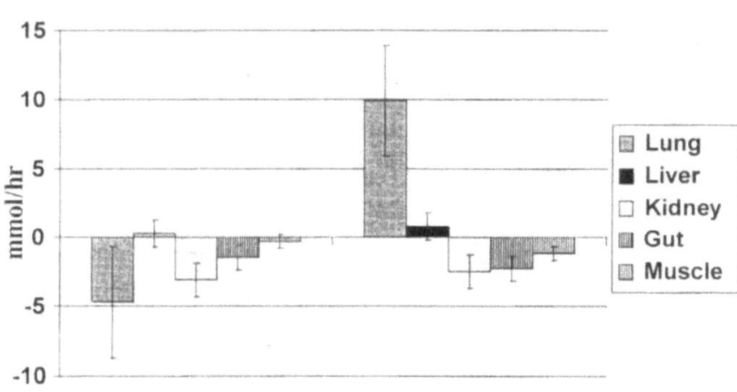

Fig. 2. Transvisceral lactate fluxes: means (mmol/h) with standard error bars. The only change from baseline to endotoxemia was for the lung ($p<0.05$)

and established adult respiratory distress syndrome (ARDS) [9]. These data suggest a scenario in which the lung is a primary producer of lactate and the gut and kidneys continue to clear it despite a drop in regional perfusion (and therefore oxygen delivery). However, the ability of these organs to clear lactate becomes impaired by decreasing organ flow and eventually is inadequate to compensate for the increased lactate production.

Data from Gutierrez et al. [9] also suggest the possibility that the lung lactate gradient reflects acute lung injury. We chose to look at this question in six patients with early ARDS and hyperlacticemia and six patients with normal arterial lactate concentrations and without ARDS [10]. The ARDS/hyperlacticemia group had a mean transpulmonary lactate gradient of 324 ± 280 mmol/h compared to 4.0 ± 9.9 mmol/h in the control group ($p < 0.01$). These data suggest that in sepsis the lung may be an important source of lactate production.

Tissue Hypoxia

The data presented above also call into question the conventional wisdom about the pathophysiological mechanisms involved in the hyperlacticemia of sepsis and suggest that other mechanisms may also be important. There is a widely held belief that in sepsis areas of tissue hypoxia result from defective microcirculatory flow, and that this forces anaerobic metabolism onto the cells in this region [11, 12]. In this view tissue hypoxia is the primary physiological event that results in lactate release in sepsis – a situation analogous to exercise. However, despite the intuitive appeal of this theory the evidence supporting it is lacking. With the use of modern techniques such as magnetic resonance spectroscopy and microfluorometric enzymatic methods Hotchkiss and coworkers have shown that there is no evidence of tissue hypoxia in septic rats when blood pressure is maintained [13]. Similar studies of human skeletal muscle have shown that tissue pO_2 actually increases in patients with sepsis [14].

Data from our studies provide further support for the concept that lactate release is not a consequence of tissue hypoxia [6]. At baseline no relationship existed between O_2 consumption (VO_2) and arterial lactate concentrations in the dog. However, during endotoxemia arterial lactate was positively correlated with VO_2 (not DO_2). Note that the correlation is in the positive direction, such that increased lactate levels are related to increased VO_2. Note also that this is true of exercising muscle in which lactate release is correlated with VO_2, not DO_2 [15]. Surprisingly, the same relationship existed for the gut during endotoxemia. Here gut lactate uptake was related to gut VO_2 during endotoxemia (Fig. 3). These data suggest that the hyperlacticemia of sepsis is fundamentally different from and perhaps significantly more complex than the hyperlacticemia occurring with other conditions such as exercise. This consideration limits the generalizability of data and concepts from exercise physiology to sepsis. Such concepts as oxygen debt and

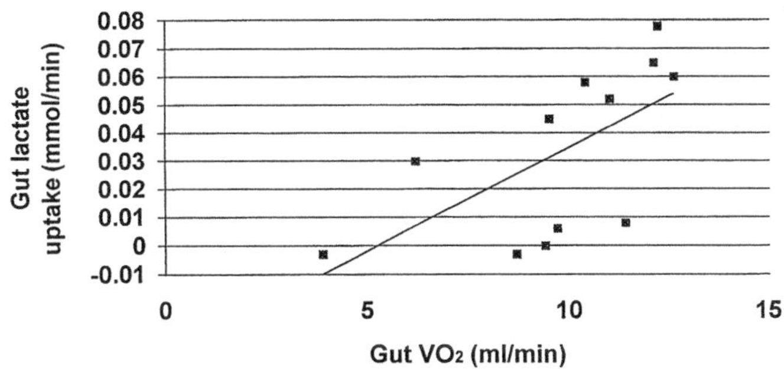

Fig. 3. Gut lactate uptake and regional oxygen consumption for the gut during the endotoxemic state (*VO_2*; $p < 0.001$). At baseline no relationship existed between gut lactate uptake and regional VO_2 (data not shown)

anaerobic threshold may not apply to patients with sepsis and hyperlacticemia.

Alternative mechanisms for hyperlacticemia have been sought. It has been known for some time that endotoxin inhibits pyruvate dehydrogenase activity in vitro [16]. This enzyme is necessary for the movement of pyruvate into the Krebs cycle and its inhibition would be expected to increase pyruvate and consequently lactate. Support for the concept that this may be important in vivo comes from human studies using the pyruvate dehydrogenase activator dichloroacetate [17–19]. These studies have failed to demonstrate any survival benefit from dichloroacetate despite decreasing lactate concentrations. Such results are not surprising if hyperlacticemia is a result of cell dysfunction rather than a cause. Such a view would hold that lactate is a marker of illness severity, but not necessarily of tissue hypoxia or ischemia.

These data are also consistent with the metabolic model that regards lactate as an obligatory product of carbohydrate metabolism to oxidative phosphorylation. Increases in blood lactate should then occur not only secondary to pyruvate dehydrogenase inhibition but also as part of the mechanisms by which injured cells improve their bioenergetics at a time of stress. This view is supported by work in isolated muscle preparations [15, 20, 21] which shows intracellular lactate accumulation in the absence of O_2 limited respiration, with the accumulation being proportional to VO_2 and the cytosolic redox state [22]. These studies suggest that the aerobic increase in cell lactate and the associated changes in NADH/NAD can influence mitochondrial NADH/NAD to support oxidative phosphorylation as the dominant source of ATP.

Other Sources of Acid

The conventional view of metabolic acidosis in sepsis relies on another important assumption. This is that lactic acid accounts for all or even most of

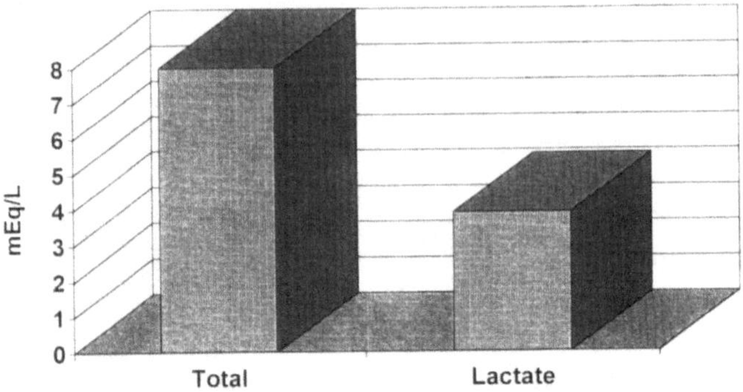

Fig. 4. Etiology of acid load in sepsis. The total acid load compared to the load from lactate acid alone is shown. Less than 50% of the total load is accounted for by lactate. (Adapted from [27])

the acidosis seen. In experimental studies of shock produced by endotoxin injection lactate accounts for less than half of the acidosis encountered [23, 24]. Similarly, in septic patients an increased anion gap has been observed that is out of proportion to the lactic acid concentration [25, 26]. In a prospective study of 30 septic patients 24 demonstrated a positive anion gap acidosis that was unaccounted for by lactate, ketones, urate, phosphate, or proteins [27]. Furthermore, in these patients with sepsis lactate accounted for less than 50% of the total metabolic acid load seen (Fig. 4). Clearly, other forms of acidosis must also be present.

Various other mechanisms leading to metabolic acidosis have been proposed: (a) other anions, besides lactate, are well known to cause acidosis in various clinical conditions (ketoacidosis, renal failure, salicylates, etc.); (b) other forms of metabolic acidosis (nonanion gap) might also occur secondary to sepsis or its treatment; (c) failure of normal homeostatic mechanisms (especially the liver); and (d) unreversed ATP hydrolysis.

We have been exploring these avenues in an effort to understand the mechanisms of acidosis in sepsis. In particular we have looked at rising anion concentrations as a potential etiology for the evolving acidosis seen in endotoxemia. However, to do this, better tools are required. The anion gap is clearly inadequate to study transorgan ion fluxes, and we have therefore turned to quantitative physical chemical techniques.

Strong Ion Gap

Stewart [28] reviewed the physical chemical properties of biological solutions over a decade ago. Although this work has been slow to move into the mainstream of biomedical research, its implications are far reaching and extremely important to our understanding of acid-base regulation in health and disease. The principles of physical chemistry define the relationship between pH and the three independent variables that control it in biological solutions. These independent variables are CO_2, the weak acids (protein and phosphate), and the strong ion difference (SID). The SID is defined as the difference between all *strong cations* minus all *strong anions*. Here, a strong ion is defined as an ion that is completely or nearly completely dissociated in water. Na^+, Cl^-, K^+, Mg^{2+}, and Ca^{2+} are all strong ions. Therefore for practical purposes SID can be defined as $(Na^+ + K^+ + Mg^{2+} + Ca^{2+}) - Cl^-$. We refer to this as the apparent SID (SIDa). For humans, the normal SIDa is 40–42 mEq/l. According to Stewart [28], if the concentrations of CO_2 and the weak acids are held constant, pH is determined by the SID. In this way differential changes in Na^+ and Cl^-, for instance, alter the pH predictably if both of the other variables are known.

A notable disadvantage of this approach was the inability precisely to quantify the acid-base impact of serum proteins. However, through the use of magnetic resonance techniques, Figge and coworkers were able to determine the charge of albumin (at various pH) and also determined that normal globulins are not significant in terms of acid-base balance [29]. Figge and coworkers also defined the exact equations for the charges on CO_2 and phosphate. They defined effective SID (SIDe) as the sum of charges for CO_2, albumin, and phosphate.

An important consequence of the work by Stewart [28] and Figge et al. [29] is that SIDa and SIDe should be exactly equal and opposite. Any "gap" between SIDa and SIDe represents unmeasured anions; or if the "gap" is negative, unmeasured cations. Lactate and urate can be measured and then excluded from this gap leaving any remaining anions or cations unexplained. Therefore, the sum of all remaining ions is equal to the size of the remaining gap. We have chosen to refer to this as the strong ion gap (SIG) [30] to distinguish it from the anion gap (Fig. 5). Normally, the SIG should be zero. When it is greater than zero, unmeasured anions outnumber unmeasured cations in the circulation. When it is less than zero, an excess of unmeasured cations exists.

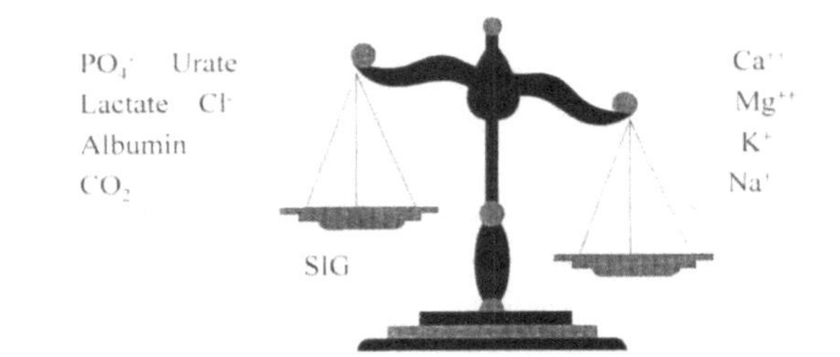

Fig. 5. Relationship between SIG and total ionic balance. *Left*, total anions; *right*, total cations. The difference between these is the strong ion gap which is shown "hanging in the balance"

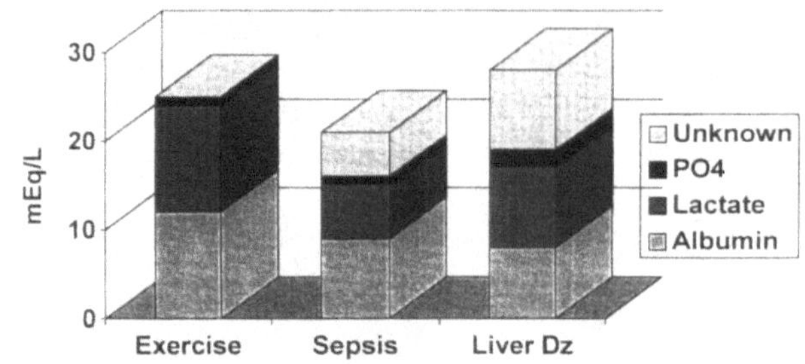

Fig. 6. Anion composition in various populations. The contributions of various anions to the total anion gap across different subject types. (Data from: exercise [46]; sepsis [27]; liver dispase [31])

The following are simplified versions of the Stewart and Figge equations:

$$SIDa = Na^{+} + K^{+} + Mg^{2+} + Ca^{2+} - Cl^{-}$$

$$SIDe \approx CO_2, \text{albumin}, PO_4$$

$$SIG = SIDa - SIDe - \text{lactate} - \text{urate}$$

When these techniques are applied to various populations (Fig. 6) we appreciate that although the anion gaps may be very similar, the compositions of these gaps vary greatly across patient types. For instance, normal subjects with lactic acidosis induced by exercise have virtually no detectable SIG, whereas patients with similar anion gaps from sepsis have SIGs which account for roughly a third of the total anion gap.

Ion Flux During Acute Endotoxemia

Unexplained anions have been measured both in patients with sepsis [27] and in those with liver disease [31]. These anions would also be expected to produce metabolic acidosis. We became intrigued by the presence of these high levels of SIG in patients with end-stage liver disease (Fig. 6). This led us to explore the liver and gut as potential sources of unexplained anions in septic shock.

Using a canine model of early endotoxemia, we sought to determine the role which the liver plays in regulating the anion/cation balance during both stable control conditions and acute endotoxemia [32]. When we examined the transorgan ion flux, a potential role emerged for the gut and liver in the regulation of blood anions. During control conditions the livers of all ten animals cleared anions from the circulation (mean flux −0.34 mEq/min). With early endotoxemia, however, the liver switched to release of anions

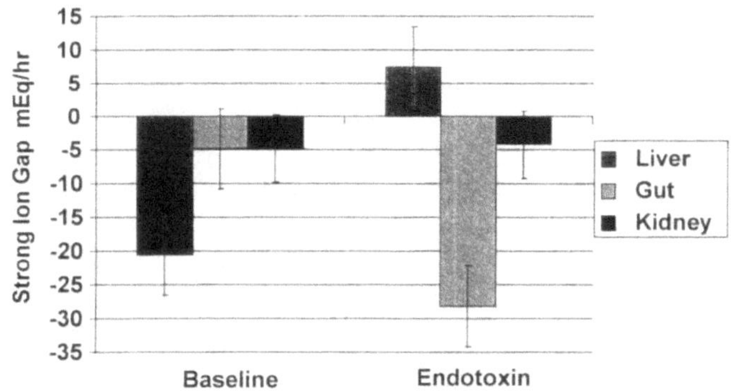

Fig. 7. Change in anion flux with endotoxin: mean total flux by hour for the various organs studied, with standard error bars. Only the liver and gut affect the blood concentrations of unexplained anions ($p<0.01$). The lung was also studied but did not effect anion concentrations (data not shown). Both the liver and gut changed their handling of anions significantly from baseline to endotoxemia ($p<0.01$)

(0.12 mEq/min, $p=0.0046$; Fig. 7). In contrast to the liver, the gut behaved in an opposite fashion. After endotoxin administration, the gut, which was neutral at baseline, began to take up anions (−0.47 mEq/min, $p=0.008$). Anion flux across the lung and kidney was unchanged. These data suggest that the gut and liver have a role in systemic acid-base balance by way of regulating anion fluxes. Interestingly, the liver did not change its handling of lactate in response to endotoxemia, and no relationship existed between transhepatic lactate and anion flux [32].

The results of this study suggest an alternative explanation for the acidosis seen during acute endotoxemia. The switch by the liver from uptake to release of anions during endotoxemia is important because any increase in anions must produce an increase in the hydrogen ion concentration. This is because any decrease in the SID results in acidosis [28]. This explanation is attractive because it requires no additional sources of hydrogen (such as ATP hydrolysis), and it fits with clinical observations made in both sepsis [27, 33, 34] and liver disease [31].

The gut also appears to play a role in anion balance, and its action is opposite to the liver. However, the increase in anion clearance by the gut did not quite offset the increased release by the liver. The total change in anion uptake for these two organs between baseline and endotoxemic conditions was 4.2 mEq/h [32]. Thus, in a 20-kg dog, the SIG would be expected to increase by 0.35 mEq/l each hour if the process were sustained. Such an increase in SIG is similar to that observed in patients with sepsis and corresponds to a decrease in pH of approximately 0.01/h (unpublished observation).

The Effect of Resuscitation on Acid-Base Balance

It has been recognized for many years that administration of large volumes of fluids, particularly 0.9% saline, produces a nonanion gap metabolic acidosis [35]. Clinically this disorder has been referred to as "dilutional" or "rehydration" acidosis [36]. The reason for this phenomenon is obvious if one considers the SID of saline, which is zero. Administration of such fluid causes a disproportionate change in the Na^+ and Cl^- concentrations in the blood of the recipient. These changes remain small unless massive volumes are given. However, in some patients with reduced "buffer capacity" from hypoproteinemia, hypophosphatemia, and similar conditions as little as 5 l of 0.9% saline would be expected to lower the pH by as much as 0.1 pH. Thus, in such a patient who suffers an insult that produces a drop in pH from 7.4 to 7.2, resuscitation with 5 l of 0.9% saline might further lower the pH to 7.1 (assuming no change in CO_2). In this case a full third of the acidosis would be coming from the saline.

Some authors, citing *in vitro* studies, have suggested that dilutional acidosis is at best only a minor issue [37]. However, preliminary data from our own *in vivo* animal studies suggest that not only is this phenomenon important, but it may actually be exaggerated in sepsis secondary to a diffuse capillary leak.

Other Potential Causes of Metabolic Acidosis

The failure of normal homeostatic mechanisms is another potential explanation for metabolic acidosis seen in sepsis. Here too we can consider the role of the liver. It is not generally appreciated that the liver has an effective mechanism for the regulation of pH. This mechanism is urea synthesis. Under normal conditions it is bicarbonate disposal that is important because protein catabolism generates large amounts of bicarbonate [38]. In acidosis urea synthesis is decreased by the liver relative to protein catabolism. This provides for a sparing effect on total body bicarbonate which would normally be consumed in the synthesis of urea [1]. At the same time, the excess nitrogen is disposed of by increased glutamine synthesis by the liver. This in turn is utilized by the kidney to effect net Cl^- ion excretion [39]. This effect would be expected to reduce extracellular Cl^- concentrations and thereby increase SID and increase pH. Interestingly, several studies have found decreased glutamine concentrations in the blood of septic animals [40, 41]. Although most authors have proposed that increased tissue utilization is the cause, preliminary data from our laboratory suggest that glutamine synthesis by the liver may also be impaired during sepsis. Further study is needed to clarify these issues. However, it is interesting to speculate on what effect this derangement in the liver's ability to respond to acidosis might have on systemic acid-base balance.

"Unreversed ATP hydrolysis" is yet another proposed mechanism for the development of acidosis in sepsis. It has been proposed that the hypoxic cell continues to hydrolyze ATP, and that the hydrogen ions liberated are released into the circulation [42]. This explanation is unsatisfactory because cellular ATP depletion does not appear to occur in experimental animal models of sepsis [43]. In human studies tissue pO_2 actually increases with worsening sepsis, further challenging the notion that cellular hypoxia is the major pathophysiological abnormality in sepsis [14] and making unreversed ATP hydrolysis an unlikely explanation for the acidosis.

Visceral Fixed Acid Uptake

In the standard nomenclature acid-base disorders are divided into metabolic and respiratory. This terminology becomes confusing, however, when it is applied to the acid itself. Both volatile acid (CO_2) and nonvolatile or "fixed" acid comes from the metabolism of substrate, and therefore both types of acid are essentially metabolic. Similarly, there is no respiratory acid as respiration does not produce acid. The condition of respiratory acidosis refers to the failure of elimination of volatile acid, whereas the condition of metabolic acidosis refers to either increased production or decreased elimination of fixed acid. With the terminology clarified, let us now consider how each organ deals with fixed acid during sepsis.

Studies using intact animals with endotoxemia allow for the quantification of acid handling by various organs. During acute endotoxemia in the dog there is a marked difference in the way various organs respond to acidemia [44] (Fig. 8). At baseline the kidney clears fixed acid from the circulation and continues to do so during endotoxemia. The liver, on the other hand, is neutral to fixed acid in both conditions. Surprisingly, the gut is not the source of

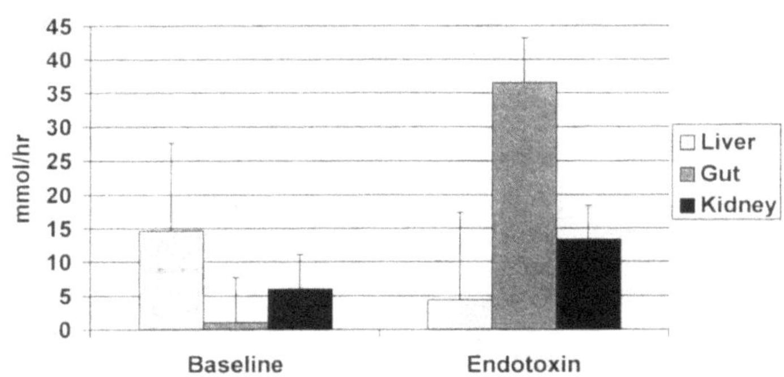

Fig. 8. Visceral fixed acid uptake: means (mmol/h) with standard error bars. Preliminary data on fixed acid flux in endotoxemia in the dog. Data from the first 30 min of endotoxemia in the unresuscitated condition. Only the gut showed a change from baseline to endotoxemia ($p<0.05$)

fixed acid. In fact, while at baseline the gut is neutral, during endotoxemia the gut actually takes up fixed acid.

What relevance to clinical medicine might such findings have? If the gut is taking up acid in conditions which are known to be associated with tissue acidosis, might the two phenomena be linked? Salzman and collegues [45] have recently shown that systemic acidosis results in intestinal mucosal acidosis regardless of the cause. In this study mucosal acidosis increased intestinal permeability even in the absence of tissue ischemia or hypoxia. Thus, during conditions of systemic acidemia the development of intestinal acidosis may result from fixed acid uptake by the gut. Such findings underscore the importance of systemic acid-base balance to intestinal mucosal integrity.

Conclusion

Patients with critical illness develop metabolic acidosis as a result of multiple processes – not just hyperlacticemia. An important source of lactate in sepsis appears to be the lung, and this may represent acute lung injury. It is perhaps just as surprising that the gut clears lactate during endotoxemia despite falling DO_2 and gut uptake of lactate is related to VO_2, not DO_2. Such data challenge the notion that hypoxia is the fundamental physiological derangement leading to hyperlacticemia and acidosis.

The gut also appears to take up other anions in sepsis as well. The liver, on the other hand, appears to release them. One effect of metabolic acidosis may be to produce tissue acidosis in the gut, and this may be a direct result of the gut's tendency to take up fixed acid. This gut acidosis may be very important because it has been shown to correlate with increased intestinal permeability even in the absence of tissue ischemia or hypoxia.

Metabolic acid flux in endotoxemia and presumably sepsis is determined by regional differences. The contributions of the splanchnic organs are complex and cannot be explained by simple, universal processes such as tissue hypoxia.

References

1. Bourke E, Häussinger D (1992) pH homeostasis: the conceptual change. In: Berlyne GM (ed) The kidney today. Selected topics in renal science. Contributions to nephrology, vol 100. Basel, Karger, pp 58–88
2. Madias NE (1986) Lactic acidosis. Kidney Int 29:752–774
3. Mizock BA, Falk JL (1992) Lactic acidosis in critical illness. Crit Care Med 20:80–93
4. van Lambalgen AA, Runge HC, van den Bos GC, Thijs LG (1988) Regional lactate production in early canine endotoxin shock. Am J Physiol 254:E45–E51
5. Cain SM, Curtis SE (1991) Systemic and regional oxygen uptake and delivery and lactate flux in endotoxic dogs infused with dopexamine. Crit Care Med 19:1552–1560
6. Bellomo R, Kellum J, Pinsky MR (1994) Visceral lactate fluxes during early endotoxemia in the dog. Am J Respir Crit Care Med 149:A413 (abstract)

7. Sayeed MM (1982) Pulmonary cellular dysfunction in endotoxin shock: metabolic and transport derangements. Circ Shock 9:335–355
8. Bowles SA, Schlichtig R, Kramer DJ, Klions HA (1992) Arteriovenous pH and partial pressure of carbon dioxide detect critical oxygen delivery during progressive hemorrhage in dogs. J Crit Care 7:95–105
9. Gutierrez G, Clark C, Nelson C, Tiu A, Brown S (1993) The lung as a source of lactate in sepsis and ARDS. Chest 104:S12 (abstract)
10. Kellum JA, Kramer DJ, Mankad S, Pinsky MR, Bellomo R, Lee K (1995) Release of lactate by the lung in acute ARDS. Crit Care Med 23:A107 (abstract)
11. Wolf YG, Cotev S, Perel A, Manny J (1987) Dependence of oxygen consumption on cardiac output in sepsis. Crit Care Med 15:198–203
12. Dantzker D (1989) Oxygen delivery and utilization in sepsis. Crit Care Clin 5:81–98
13. Hotchkiss RS, Rust RS, Dence CS et al (1991) Evaluation of the role of cellular hypoxia in sepsis by hypoxic marker [^{18}F] fluoromisonidazole. Am J Physiol 261:R965–R972
14. Boekstegers P, Weidenhofer S, Kapsner T, Werdan K (1994) Skeletal muscle partial pressure of oxygen in patients with sepsis. Crit Care Med 22:640–650
15. Connett RJ, Gayeski TEJ, Honig CR (1986) Lactate efflux is unrelated to intracellular PO_2 in working red muscle in situ. J Appl Physiol 61:402–406
16. Kilpatrick-Smith L, Dean J, Erecinska M, Silver IA (1983) Cellular effects of endotoxin in vitro. II. Reversibility of endotoxic damage. Circ Shock 11:101–111
17. Stacpoole PW, Harman EM, Curry SH, Baumgartner TJ, Misbin RI (1983) Treatment of lactic acidosis with dichloroacetate. N Engl J Med 309:390–396
18. Preiser JC, Moulart D, Vincent J-L (1990) Dichloroacetate administration in the treatment of endotoxin shock. Circ Shock 30:221–228
19. Curtis SE, Cain SM (1992) Regional and systemic oxygen delivery/uptake relations and lactate flux in hyperdynamic endotoxin-treated dogs. Am Rev Respir Dis 145:348–354
20. Connett RJ, Gayeski TEJ, Honig CR (1984) Lactate accumulation in fully aerobic, working, dog gracilis muscle. Am J Physiol 246:H120–H128
21. Connett RJ, Gayeski TEJ, Honig CR (1985) Energy sources in fully aerobic rest-work transitions: a new role for glycolysis. Am J Physiol 248:H922–H929
22. Connett RJ, Honig CR, Gayeski TEJ, Brooks GA (1990) Defining hypoxia: a systems view of VO_2, glycolysis, energetics and intracellular PO_2. J Appl Physiol 68:833–842
23. van Lambalgen AA, Bronsveld W, van den Bos GC, Thijs LG (1984) Distribution of cardiac output, oxygen consumption and lactate production in canine endotoxin shock. Cardiovasc Res 18:195–205
24. Rackow EC, Mecher C, Astiz ME, Goldstien C, McKee D, and Weil MH (1990) Unmeasured anion during severe sepsis with metabolic acidosis. Circ Shock 30:107–115
25. Gabow P, Kaehny W, Fennessey P, Goodman SI, Gross PA, Schrier RW (1980) Diagnostic importance of an increased serum anion gap. N Engl J Med 303:854–858
26. Mehta K, Kruse JA, Carlson RW (1986) The relationship between anion gap and elevated lactate. Crit Care Med 14:405
27. Mecher C, Rackow EC, Astiz ME, Weil MH (1991) Unaccounted for anion in metabolic acidosis during severe sepsis in humans. Crit Care Med 19:705–711
28. Stewart P (1981) How to understand acid-base: a quantitative acid-base primer for biology and medicine. Elsevier, New York
29. Figge J, Mydosh T, Fencl V (1992) Serum proteins and acid-base equilibria: a follow-up. J Lab Clin Med 120:713–719
30. Kellum JA, Kramer DJ, Pinsky MR (1995) Strong ion gap: a methodology for exploring unexplained Anions. J Crit Care 10:51–55
31. Kellum JA, Kramer DJ, Pinsky MR (1994) Unexplained positive anion gap metabolic acidosis in end stage liver disease. Crit Care Med 22:A209 (abstract)
32. Kellum JA, Bellomo R, Kramer DJ, Pinksy MR (1995) Hepatic anion flux during acute endotoxemia. J Appl Physiol 78:2212–2217

33. Rackow EC, Mecher C, Astiz ME, Goldstien C, McKee D, Weil MH (1990). Unmeasured anion during severe sepsis with metabolic acidosis. Circ Shock 30(2):107–115
34. Gilfix BM, Bique M, Magder S (1993) A physical chemical approach to the analysis of acid-base balance in the clinical setting. J Crit Care 8(4):187–197
35. Cheek DB (1956) Changes in total chloride and acid-base balance in gastroenteritis following treatment with large and small loads of sodium chloride. Pediatrics 17:839–847
36. Narins RG, Jones ER, Townsend R, Goodkin DA, Shay RJ (1985) Metabolic acid-base disorders: pathophysiology, classification and treatment. In: Arieff AI, DeFronzo RA (eds) Fluid, electrolyte, and acid-base disorders. Churchill Livingstone, New York, pp 269–384
37. Garella S, Chang BS, Kahn SI (1975) Dilution acidosis and contraction alkalosis: review of a concept. Kidney Int 8:279–283
38. Atkinson DE, Camien MN (1982) The role of urea synthesis in removal of metabolic bicarbonate and regulation of blood pH. Curr Top Cell Regul 21:261–302
39. Häussinger D, Gerok W, Sies H (1986) The effect of urea synthesis on extracellular pH in the isolated perfused rate liver. Biochem J 236:261–265
40. Parry-Billings M, Leighton B, Dimitriadis G, de Vasconcelos PRL, Newssholme EA (1986) Skelatal muscle glutamine metabolism during sepsis in the rat. Int J Biochem 21:419–423
41. Higashiguchi T, Hasselgren PO, Nogucho Y, James JH, Fischer JE (1992) Altered metabolism of glutamine by isolated enterocyetes during sepsis. Society for Surgery of the Alimentary Tract, 33rd annual meeting, p 101 (abstract)
42. Gores GJ, Flarsheim CE, Dawson TL, Nieminen A-L, Herman B, Lemasters JJ (1989) Swelling, reductive stress, and cell death during chemical hypoxia in hepatocytes. Am J Physiol 26:C347–C354
43. Jepson MM, Cox M, Bates PC et al (1987) Regional blood flow and skelatal muscle energy status in endotoxemic rats. Am J Physiol 252:E581–E587
44. Kellum JA, Bellomo R, Kramer DJ, Pinsky MR (1995) Fixed acid uptake by visceral organs during early endotoxemia. Crit Care Med 23:A173 (abstract)
45. Salzman AL, Wang H, Wollert PS et al (1994) Endotoxin-induced ileal mucosal hyperpermeability in pigs: role of tissue acidosis. Am J Physiol 266:G633–G646
46. Lindinger MI, Heigenhauser GJF, McKelvie RS, Jones NL (1992) Blood ion regulation during repeated maximal exercise and recovery in humans. Am J Physiol 262:R126–R136

The Splanchnic Microcirculation: Cytokines and Cell Adhesion Molecules

K. Ley

Introduction

This chapter focuses on the function of the intestinal microcirculation in the inflammatory response, which is characterized by leukocyte infiltration, increased microvascular permeability, edema, and frequently also tissue damage. Leukocyte traffic in inflammation is regulated by chemoattractants and by specific adhesion molecules, whose expression on endothelial cells is controlled by cytokines. Regulation of cell adhesion at the blood-endothelial interface has received considerable attention in the past decade, which has led to the development of a conceptual framework for leukocyte movement in inflammatory and immune responses. The recent availability of specific antibodies, peptides, carbohydrates, and gene-deficient and transgenic rodents is critical for further elucidating the mechanisms of intestinal inflammation.

Splanchnic Microcirculation

The intestinal microcirculation is supplied by the celiac, superior mesenteric, and inferior mesenteric arteries and drained by the portal vein. Blood flow in the small intestine is of the order of 40–80 ml/min for 100 g tissue and accounts for 10%–15% of cardiac output at rest [2]. Numerous arcades at different levels of the circulation serve to ensure a constant feeding pressure into the microcirculation along the gut. The larger arcading vessels represent a site of resistance to flow, leading to an inflow pressure into arterioles with 50 mm diameter of 40–50 mmHg [3]. The submucosal microvascular plexus sends separate branches to the muscularis and mucosal layers. Mucosal blood flow is regulated mainly by precapillary resistance, causing capillary pressure to be very close to portal venous outflow pressure (approximately 10 mmHg). This arrangement makes the mucosal tissue susceptible to edema formation if venous outflow pressure is elevated and/or microvascular permeability is increased. Indeed, significant increases of colloid osmotic pressure in plasma and reductions in the interstitial fluid have been reported in the cat small intestine upon elevation of venous outflow pressure [4]. The distribution of microvascular parameters including diameters, blood flow velocities, hematocrit, and leukocyte concentration shows a large degree of spatial heterogeneity, as concluded from analysis of the mesenteric microcirculation [5–7]. This heterogeneity includes the formation of functional arteriovenous shunts which can be detected by microsphere and clearance measurements but do not appear to have a clear anatomical correlate. The mucosal capillary network is lined by a fenestrated endothelium, while the muscularis is supplied by capillaries with continuous endothelium [2]. In addition, specialized "high" endothelial cells are present in mucosal lymphoid tissues, for example, Peyer's patches [8, 9].

Intestinal Inflammation

The intestines are the site of a variety of clinically relevant inflammatory disorders including inflammatory bowel disease (Crohn's disease and ulcerative colitis), inflammation induced by nonsteroidal anti-inflammatory drugs, gastro-duodenitis induced by *Helicobacter pylori* infection, and the inflammatory response following intestinal ischemia and reperfusion. The intestinal microcirculation is the source of cytokines and other mediators leading to damage in the lung [10] or liver [11]. Life-threatening systemic diseases such as sepsis and endotoxin shock can result from the breakdown of the epithelial barrier between the intestinal lumen and the blood compartment.

Experimentally, inflammation in intestinal tissues can be induced by surgical trauma [12, 13], by local application of chemoattractants or cytokines [14–16], by generation of oxygen-derived free radicals by enzymatic systems, or by ischemia followed by reperfusion [17, 18]. Several experimental interventions are known to produce intestinal inflammation resembling ulcerative colitis or Crohn's disease [19], including the introduction of defects in the genes encoding for interleukin (IL)-2, IL-10, or components of the T-cell receptor [20].

Endothelial Activation

Endothelial cells respond in a complex way to inflammatory stimuli by elaborating cytokines, chemoattractants, arachidonic acid metabolites, adhesion molecules, and other surface receptors. By virtue of this array of activities the microvascular endothelium is of paramount importance to the regulation of the inflammatory response. Cultured endothelial cells respond to mechanical, chemical, or ischemic injury by producing IL-6, granulocyte-macrophage colony stimulating factor (GM-CSF), and other cytokines, by changing their pattern of enzymatic activity, and by expressing major histocompatibility complex-II (MHC II), urokinase, and adhesion molecules [21]. At least two different time courses of activation can be distinguished readily. Certain inflammatory mediators such as histamine or thrombin cause immediate exocytosis of endothelial cell granules containing preformed molecules, for example, von Willebrand factor and P-selectin [22].

Cytokines and Endothelium

Very little is known about the cellular effects of cytokines on microvascular endothelium in situ, but cultured endothelial cells have been used to investigate the basic mechanisms, which are likely to be operative in the microcirculation as well (Table 1). Cytokine-mediated activation of endothelial cells usually initiates protein synthesis dependent expression of secreted and/or surface molecules [23]. Tumor necrosis factor-α (TNF-α) and IL-1 bind to specific cell surface receptors and initiate signal transduction cascades which are not fully understood at this time. In addition to activating intracellular protein kinases, TNF-α appears to release ceramide from cell membrane sphingomyelin [24], which acts as a second messenger. Similar signal transduction pathways are discussed for endothelial cell activation by IL-1 [25, 26]. A common consequence of cytokine stimulation of endothelial cells is the activation of nuclear response elements such as nuclear factor kappa-B (NF-κB) [27], which result in gene transcription. Genes regulated by IL-1 and TNF that also contain NF-κB sites in their 5' flanking regions include MHC I and II, GM-CSF, IL-6, monocyte chemotactic protein-1 (MCP-1), urokinase, and several leukocyte-endothelial adhesion molecules [21]. Unlike IL-1 and TNF-α, which induce transcription of a variety of genes, IL-4 appears to have a more specific effect on endothelial cells in that it induces surface expression of vascular cell adhesion molecule-1 but not other induci-

Table 1. Cytokine effects on endothelial cell function and production of cytokines by endothelial cells

Cytokine	Effect/function
Cytokines affecting endothelial cell function	
IL-1	Transcription of genes for adhesion molecules, chemoattractants, enzymes
TNF-α	Transcription of genes for adhesion molecules, chemoattractants, enzymes DNA degradation by endonucleases, induction of apoptosis
IL-4	Expression of vascular cell adhesion molecule-1 Selective enhancement of T-lymphocyte adhesion Expression of ligand(s) for leukocyte L-selectin
IFN-γ	Upregulation of MHC-I and II Synergistic with IL-1 for expression of intercellular adhesion molecule-1 [1]
Cytokines produced by endothelial cells	
IL-1	Proinflammatory functions, autocrine effects (?)
IL-6	Pleiotropic responses in liver, B-lymphocytes and other organ systems
IL-8	Chemoattractant/activator for neutrophils
M-CSF	Stimulates monocyte production
GM-CSF	Stimulates granulocyte and monocyte production, chemoattractant
MCP-1	Chemoattractant for mononuclear phagocytes

ble adhesion molecules [28]. After stimulation with interferon-γ, endothelial cells show still another pattern of activation, which includes the expression of MHC-I, MHC-II, and intercellular adhesion molecule-1, but not other adhesion molecules [23].

Microvascular endothelium is not only a target, but also a source of cytokines. Again, most of the available data are based on the study of cultured endothelial cells in vitro. Endothelial cells produce IL-8, which is a chemoattractant cytokine (chemokine) specific for neutrophils and monocytes (reviewed in [29]). IL-8 can bind to a variety of extracellular matrix glycoproteins and cell surface proteoglycans [30, 31], which may be important for the generation of a locally stable gradient of bound chemoattractant. Indeed, IL-8 has been shown to be an important endogenous chemoattractant in an in vitro model of the vascular wall [32]. On the other hand, IL-8 can also have anti-inflammatory effects if the circulation is flooded with this cytokine by intravenous injection [33, 34]. This apparent paradox may have a bearing for patients with septicemia or endotoxin shock, in which circulating chemoattractants may inhibit neutrophil recruitment to specific sites. The loss of specificity of leukocyte emigration may be related to secondary organ failure in septic patients. Other important chemoattractants/activators produced and secreted by activated endothelial cells include platelet-activating factor [35] and leukotrienes [36]. In addition to its colony-stimulating activity, endothelial-derived GM-CSF also acts as an activator/attractant for neutrophils and monocytes. MCP-1 is a chemoattractant specific for mononuclear phagocytes [37] and appears to play a role in the pathogenesis of atherosclerotic lesions [38]. Activated endothelial cells produce IL-6, which elicits multiple systemic responses including secretion of acute-phase proteins by liver cells and induction of fever [39]. IL-1 is also secreted by activated endothelium, which may be involved in autocrine stimulation during inflammation.

Adhesion Molecules

On stimulation, endothelial cells express a number of adhesion molecules which are relevant to leukocyte traffic in inflammation. These adhesion molecules can broadly be grouped into four families. *Integrins* are αβ-heterodimers which recognize extracellular matrix and cell surface glycoproteins as well as some soluble molecules such as complement factor C3bi. Integrins usually require a conformational change to gain full adhesive function [40]. Endothelial cells can express ligands for the β_2 subclass of integrins which are restricted to neutrophils, lymphocytes, and monocytes. The *selectins* are a family of three carbohydrate-recognizing transmembrane molecules, two of which, E- and P-selectin, are expressed on activated endothelium. The largest family of endothelial adhesion molecules belong to the *immunoglobulin* superfamily of glycoproteins. Some can engage in homotypic interaction, while others serve as ligands for β_1 and β_2 integrins expressed on circulating lymphocytes and granulocytes. Finally, an emerging family of *mucin-like ad-*

hesion molecules are defined by their rich glycosylation, predominantly via O-linked side chains [41]. Its members represent counterreceptors for selectins [42].

The physiology of inflammation is characterized by initial transient leukocyte adhesion to the endothelium (rolling), followed by leukocyte arrest and transmigration (for review see [43, 44]). Leukocyte rolling is mediated by L-selectin and P-selectin, while firm adhesion requires integrin-immunoglobulin interaction. For the sake of clarity, the expression, regulation, and function of individual endothelial-leukocyte adhesion molecules are presented in Table 2.

Mucosal Immune System

Specialized lymphatic organs including Peyer's patches and the mesenteric lymph node are found along the alimentary canal. Moreover, two classes of lymphocytes are resident in the intestinal wall, i.e., within the epithelium (intraepithelial lymphocytes) and in the lamina propria [93, 94]. In contrast to lymphocyte homing to peripheral lymph nodes, lymphocyte entry into Peyer's patches requires binding of $\alpha_4\beta_7$ and related integrins to the so-called mucosal addressin complex [95]. Firm lymphocyte adhesion and entry into the Peyer's patches depend on activation mediated by a G-protein coupled receptor [8]. In addition, L-selectin plays a partial role in lymphocyte homing to mucosal sites [65, 96]. The mucosal immune system is an area of intense study because it is involved in inflammatory bowel disease, autoimmune diseases, and possible intestinal tumors.

Conclusion: Potential Anti-inflammatory Therapies

Adhesion molecules and cytokines represent attractive targets for the rational design of anti-inflammatory compounds. The important role of adhesion molecules in mediating organ damage is illustrated by the recent finding that mice rendered deficient for expression of intercellular adhesion molecule-1 are resistant to experimental endotoxin shock, apparently because their livers are protected from damage mediated by activated granulocytes [97]. Although a number of preclinical studies have shown that antibodies, peptides, and carbohydrates can limit neutrophil infiltration of inflammatory sites, specific problems are commonly encountered on the way to a better anti-inflammatory therapy. To name but a few, adhesion receptors generally have fairly low affinities for substrates, and small molecule inhibitors tend to be worse than the natural ligands which requires high concentration of inhibitor to be effective. Also, the circulatory half-life of some potential inhibitors is very unsatisfactory. On the other hand, the engineering of monoclonal antibodies poses technological and economical problems, and even humanized monoclonal antibodies injected into patients are likely to elicit an immune re-

Table 2. Endothelial-leukocyte adhesion molecules

Name(s)	Expression	Regulation	Ligand(s)	Function
Integrins				
$\alpha_L\beta_2$ LFA-1 CD11a/CD18	Lymphocytes > neutrophils, monocytes	Conformationally activated by chemoattractants	ICAM-1, ICAM-2, ICAM-3	Lymphocyte, granulocyte adhesion to resting endothelium [45, 46]
$\alpha_M\beta_2$ Mac-1 CD11b/CD18	Macrophages > monocytes, neutrophils	Upregulated by degranulation activated by TNF, chemoattractants and conformationally activated	ICAM-1, C3bi, unidentified endothelial ligand	Granulocyte, monocyte adhesion to resting and activated endothelium [47, 48] Transmigration, C3bi binding [49] Leucocyte activation, phagocytosis [40]
$\alpha_X\beta_2$ p150,95 CD11c/CD18	Macrophages, monocytes	Upregulated by degranulation and conformation	C3bi, others?	Complement fixation, phagocytosis? [50, 51]
$\alpha_4\beta_1$ CD49d/CD29	Lymphocytes, monocytes, eosinophils	Constitutively active (?) affinity regulation [53], higher expression on memory T-cells	VCAM-1	Lymphocyte, monocyte adhesion to endothelium and fibronectin [52, 53]
$\alpha_4\beta_7$	Lymphocytes	Constitutive	MAdCAM-1	Mucosal lymphocyte homing [40, 54]
Selectins				
E-selectin CD62E	Activated endothelium	Up-regulated by TNF, IL-1, max. expression at 4–6 h	Sialyl-Lewis^x glycoproteins, glycolipids on PMN	Adhesion and rolling of neutrophils [55–57] Adhesion of T-cells, monocytes [58, 59] Absence has no (?) effect on neutrophil traffic in vivo [60]
L-selectin CD62L	Resting neutrophils, most lymphocytes	Constitutive, shed from surface by chemoattractant	GlyCAM-1, CD34 on lymph node HEV; evidence for extralymphatic ligand(s)	Lymphocyte homing to lymph nodes [61, 62] Sustained leukocyte rolling in vivo [15, 63, 64] Critical for PMN recruitment [65]
P-selectin CD62P	Activated platelets, thrombin-, histamine-, cytokine-activated endothelium regulation	Surface-expressed by degranulation, transcriptional regulation by TNF, IL-1	PSGL-1 on myeloid cells	Platelet-leukocyte adhesion [66–68] Early leukocyte rolling in vivo [69, 70] Critical for PMN recruitment [70, 71]

Immunoglobulins				
ICAM-1 CD54	Endothelium, smooth muscle, fibroblasts, T-lymphocytes	Constitutive, increased expression by IL-1, IFN-γ	LFA-1, Mac-1	Cytotoxic T-cell conjugates, leukocyte-endothelial adhesion, transmigration? [72, 73]
ICAM-2 CD102	Endothelial cells	Constitutive	LFA-1	Leukocyte adhesion to resting endothelium [74, 75]
ICAM-3 CD50	Resting T-cells	Constitutive	LFA-1	Homotypic lymphocyte aggregation [76, 77]
VCAM-1	Cytokine-stimulated endothelium	IL-1, IL-4, TNF up-regulate expression, max. at 4–24 h	VLA4 ($\alpha_4\beta_1$ integrin) $\alpha_4\beta_7$ integrin	Adhesion of monocytes, eosinophils, lymphocytes to activated endothelium [78, 79]. Expressed in atherosclerotic plaques [80]
PECAM-1 CD31	High in endothelium, moderate on monocytes, PMN, platelets	Constitutive redistribution to cell borders in confluent EC	PECAM-1 (homotypic) other ligand(s)?	Critically involved in neutrophil transmigration, endothelial monolayer integrity, neutrophil adhesion (?) [81, 82]
MAdCAM-1	Mucosal venules, Peyer's patch HEV, mesenteric lymph node	Constitutive, different glycosylation in Peyer's patch and mesentenic lymph node	$\alpha_4\beta_7$ integrin L-selectin (glycosylation dependent)	Lymphocyte homing to mucosal sites lymphocyte rolling (contains mucin domain) [83, 84]
Selectin ligands				
PSGL-1	Myeloid cells	Constitutive, dimer	P-selectin	Adhesion of myeloid cells [41, 85]
GlyCAM-1	Lymph node HEV, serum	Constitutive, secreted, expression depends on lymph flow	L-selectin	Homing of lymphocytes to lymph node HEV (?), modulation of L-selectin-dependent adhesion? [86, 87]
CD34	Lymph node HEV, other endothelium	Constitutive, expression depends on lymph flow	L-selectin	Homing of lymphocytes to lymph node HEV [41, 88]
ESL-1	Murine myeloid cells	Constitutive	E-selectin	Neutrophil adhesion [98]
Partially identified	Murine myeloid cells	Constitutive	P-selectin	Unknown [89, 90]
Unidentified	Activated endothelial cells	Induced by TNF in vitro, by tissue trauma in vivo	L-selectin	Mediates neutrophil adhesion [91, 92] L-selectin dependent rolling [28, 64]

sponse. Cytokine inhibitors such as IL-1 receptor antagonist have shown beneficial effects in some models, but more clinical studies are needed. The issue of specificity, i.e., the ability to attack pathologic inflammation while preserving the protection of the organism from its environment by inflammatory mechanisms, is unsolved and has seldom even been addressed yet.

The level of interest in novel anti-inflammatory therapies is demonstrated by numerous conferences organized in the field. Currently, selectin antagonists, oligonucleotide-based inhibition of endothelial adhesion molecules, and several approaches aimed at inhibiting β_2 integrins are being considered. Clearly, the current state of knowledge provides the basis for rational design of preclinical and clinical studies. Understanding the physiology of intestinal inflammation and using the insights gained to conceive better therapies represents a challenge for future studies.

References

1. Dustin ML, Rothlein R, Bhan AK, Dinarello CA, Springer TA (1986) Induction by IL-1 and interferon-gamma: Tissue distribution, biochemistry, and function of a natural adherence molecule (ICAM-1). J Immunol 137:245–254
2. Lundgren O (1984) Microcirculation of the gastrointestinal tract and pancreas. In: Renkin EM, Michel CC (eds) Handbook of Physiology vol IV (2). The cardiovascular system. American Physiological Society, Bethesda, pp 799–863
3. Gore RW, Bohlen HG (1977) Microvascular pressures in rat intestinal muscle and mucosal villi. Am J Physiol 233:H685–H693
4. Mortillaro NA, Taylor AE. (1976) Interaction of capillary and tissue forces in the cat small intestine. Circ Res 39:348–358
5. Ley K, Pries AR, Gaehtgens P (1988) Preferential distribution of leukocytes in rat mesentery microvessel networks. Pflugers Arch 412:93–100
6. Pries AR, Ley K, Gaehtgens P (1986) Generalization of the Fahraeus principle for microvessel networks. Am J Physiol 251:H1324–H1332
7. Pries AR, Secomb TW, Gaehtgens P, Gross JF (1990) Blood flow in microvascular networks: experiments and simulation. Circ Res 67:826–834
8. Bargatze RF, Butcher EC (1993) Rapid G protein-regulated activation event involved in lymphocyte binding to high endothelial venules. J Exp Med 178:367–372
9. Chin YH, Rasmussen RA, Woodruff JJ (1986) A monoclonal anti-HEBFpp antibody with specificity for lymphocyte surface molecules mediating adhesion to Peyer's patch high endothelium of the rat. J Immunol 136:2556–2561
10. Carden DL, Young JA, Granger DN (1993) Pulmonary microvascular injury after intestinal ischemia-reperfusion: role of P-selectin. J Appl Physiol 75:2529–2534
11. Jaeschke H, Farhood A, Smith CW (1991) Neutrophil-induced liver cell injury in endotoxin shock is a CD11b/CD18-dependent mechanism. Am J Physiol 261:G1051–G1056
12. Fiebig E, Ley K, Arfors K-E (1991) Rapid leukocyte accumulation by "spontaneous" rolling and adhesion in the exteriorized rabbit mesentery. Int J Microcirc Clin Exp 10:127–144
13. Ley K (1994) Histamine can induce leukocyte rolling in rat mesenteric venules. Am J Physiol 267:H1017–H1023
14. Ley K, Bullard D, Arbones ML, Bosse R, Vestweber D, Tedder TF et al (1994) Sequential contribution of L- and P-selectin to leukocyte rolling in vivo. J Exp Med 181:669–675

15. von Andrian UH, Chambers JD, McEvoy LM, Bargatze RF, Arfors K-E, Butcher EC (1991) Two step model of leukocyte-endothelial cell interaction in inflammation: distinct roles for LECAM-1 and the leukocyte β_2 integrins in vivo. Proc Natl Acad Sci USA 88:7538–7542
16. Von Andrian UH, Hansell P, Chambers JD, Berger EM, Torres Filho I, Butcher EC et al (1992) L-selectin function is required for β_2 integrin-mediated neutrophil adhesion at physiological shear rates in vivo. Am J Physiol 263:H1034–H1044
17. Grisham MB, Hernandez LA, Granger DN (1986) Xanthine oxidase and neutrophil infiltration in intestinal ischemia. Am J Physiol 251:G567–G574
18. Kubes P (1993) Ischemia-reperfusion in feline small intestine: a role for nitric oxide. Am J Physiol 264:G143–G149
19. Sartor RB (1992) Animal models of intestinal inflammation. In: McDermott RP, Stenson WF (eds) Inflammatory bowel disease. Elsevier, New York, pp 337–353
20. Strober W, Ehrhardt RO (1993) Chronic intestinal inflammation: an unexpected outcome in cytokine or T cell receptor mutant mice. Cell 75:203–205
21. Gerritsen ME, Bloor CM (1993) Endothelial cell gene expression in response to injury. FASEB J 7:523–532
22. Hattori R, Hamilton KK, Fugate RD, McEver RP, Sims PJ (1989) Stimulated secretion of endothelial von Willebrand factor is accompanied by rapid redistribution to the cell surface of the intracellular granule membrane protein GMP-140. J Biol Chem 264:7768–7771
23. Pober JS, Cotran RS (1990) Cytokines and endothelial cell biology. Physiol Rev 70:427–451
24. Schütze S, Potthoff K, Machleidt T, Berkovic D, Wiegmann K, Krönke M (1992) TNF activates NF-kappa B by phosphatidylcholine-specific phospholipase C-induced "acidic" sphingomyelin breakdown. Cell 71:765–776
25. Kolesnick R, Golde DW (1994) The sphingomyelin pathway in tumor necrosis factor and interleukin-1 signaling. Cell 77:325–328
26. Schütze S, Machleidt T, Krönke M (1994) The role of diacylglycerol and ceramide in tumor necrosis factor and interleukin-1 signal transduction. J Leukocyte Biol 56:533–541
27. Collins T (1993) Endothelial nuclear factor-kappa B and the initiation of the atherosclerotic lesion. Lab Invest 68:499–508
28. Luscinskas FW, Kansas GS, Ding H, Pizcueta P, Schleiffenbaum B, Tedder TF et al (1994) Monocyte rolling, arrest and spreading on IL-4-activated vascular endothelium under flow is mediated via sequential action of L-selectin, β_1–integrins, and β_2–integrins. J Immunol 125:1417–1427
29. Baggiolini M (1993) Novel aspects of inflammation: interleukin-8 and related chemotactic cytokines. Clin Invest 71:812–814
30. Rot A (1992) Endothelial cell binding of NAP-1/IL-8:role in neutrophil emigration. Immunol Today 13:291–294
31. Tanaka Y, Adams DH, Shaw S (1993) Proteoglycans on endothelial cells present adhesion-inducing cytokines to leukocytes. Immunol Today 14:111–115
32. Huber AR, Kunkel SL, Todd RF, Weiss SJ (1991) Regulation of transendothelial neutrophil migration by endogenous interleukin-8. Science 254:99–102
33. Hechtman DH, Cybulsky MI, Fuchs HJ, Baker JB, Gimbrone MAJ (1991) Intravascular interleukin-8:an inhibitor of polymorphonuclear leukocyte accumulation at sites of acute inflammation. J Immunol 147:883–892
34. Ley K, Baker JB, Cybulsky MI, Gimbrone MAJ, Luscinskas FW (1993) Intravenous interleukin-8 inhibits granulocyte emigration from rabbit mesenteric venules without altering L-selectin expression or leukocyte rolling. J Immunol 151:6347–6357
35. McIntyre TM, Zimmerman GA, Prescott SM (1986) Leukotrienes C4 and D4 stimulate human endothelial cells to synthesize platelet-activating factor and bind neutrophils. Proc Natl Acad Sci USA 83:2204–2208
36. Feinmark SJ (1992) The role of the endothelial cell in leukotriene biosynthesis. Am Rev Respir Dis 146:S51–S55

37. Wang JM, Sica A, Peri G, Walter S, Padura IM, Libby P et al (1991) Expression of monocyte chemotactic protein and interleukin-8 by cytokine-activated human vascular smooth muscle cells. Arterioscler Thromb 11:1166–1174
38. Miller MD, Krangel MS (1992) Biology and biochemistry of the chemokines: a family of chemotactic and inflammatory cytokines. Crit Rev Immunol 12:17–46
39. Fey GH, Hattori M, Hocke G, Brechner T, Baffet G, Baumann M et al (1991) Gene regulation by interleukin 6. Biochimie 73:47–50
40. Hynes RO (1992) Integrins: versatility, modulation, and signaling in cell adhesion. Cell 69:11–25
41. Shimizu Y, Shaw S (1993) Mucins in the mainstream. Nature 366:630–631
42. Rosen SD (1993) Ligands for L-selectin – where and how many? Res Immunol 144:699–703
43. Ley K (1989) Granulocyte adhesion to microvascular and cultured endothelium. Stud Biophys 134:179–184
44. Ley K (1992) Leukocyte adhesion to vascular endothelium. J Reconstr Microsurg 8:495–503
45. Springer TA (1990) Adhesion receptors of the immune system. Nature 346:425–434
46. Springer TA, Thompson WS, Miller LJ, Schmalstieg FC, Anderson DC (1984) Inherited deficiency of the Mac-1, LFA-1, p150,95 glycoprotein family and its molecular basis. J Exp Med 160:1901–1918
47. Corbi AL, Kishimoto TK, Miller LJ, Springer TA (1988) The human leukocyte adhesion glycoprotein Mac-1 (complement receptor type 3, CD 11b) α subunit. Cloning, primary structure and relation to the integrins, von Willebrand factor and factor B. J Biol Chem 263:12403–12411
48. Diamond MS, Springer TA (1993) A subpopulation of Mac-1 (CD11b/CD18) molecules mediates neutrophil adhesion to ICAM-1 and fibrinogen. J Cell Biol 120:545–556
49. Wright SD, Rao PE, Van Voorhis WC, Craigmyle LS, Iida K, Talle MA et al (1983) Identification of the C3bi receptor of human monocytes and macrophages by using monoclonal antibodies. Proc Natl Acad Sci USA 80:5699–5703
50. Corbi AL, Miller LJ, O'Connor K, Larson RS, Springer TA (1987) cDNA cloning and complete primary structure of the α subunit of a leukocyte adhesion glycoprotein p150,95. EMBO J 6:4023–4028
51. Keizer GD, te Velde AA, Schwarting R, Figdor CG, de Vries JE (1987) Role of p150,95 in adhesion, migration, chemotaxis and phagocytosis of human monocytes. Eur J Immunol 17:1317–1322
52. Dobrina A, Menegazzi R, Carlos TM, Nardon E, Cramer R, Zacchi T et al (1991) Mechanisms of eosinophil adherence to cultured vascular endothelial cells. Eosinophils bind to the cytokine-induced endothelial ligand vascular cell adhesion molecule-1 via the very late activation antigen-4 integrin receptor. J Clin Invest 88:20–26
53. Shimizu Y, van Seventer GA, Horgan KJ, Shaw S (1990) Regulated expression and binding of three VLA (β_1) integrin receptors on T cells. Nature 345:250–253
54. Holzmann B, Weissman IL (1989) Integrin molecules involved in lymphocyte homing to Peyer's patches. Immunol Rev 108:45–60
55. Bevilacqua MP, Pober JS, Mendrick DL, Cotran RS, Gimbrone MAJ (1987) Identification of an inducible endothelial-leukocyte adhesion molecule. Proc Natl Acad Sci USA 84:9238–9242
56. Bevilacqua MP, Stengelin S, Gimbrone MAJ, Seed B (1989) Endothelial leukocyte adhesion molecule-1: an inducible receptor for neutrophils related to complement regulatory proteins and lectins. Science 243:1160–1165
57. Lawrence MB, Bainton DF, Springer TA (1994) Neutrophil tethering to and rolling on E-selectin are separable by requirement for L-selectin. Immunity 1:137
58. Carlos T, Kovach N, Schwartz B, Rosa M, Newman B, Wayner E et al (1991) Human monocytes bind to two cytokine-induced adhesive ligands on cultured human endothelial cells – endothelial-leukocyte adhesion molecule-1 and vascular cell adhesion molecule-1. Blood 77:2266–2271

59. Picker LJ, Kishimoto TK, Smith CW, Warnock RA, Butcher EC (1991) ELAM-1 is an adhesion molecule for skin-homing T cells. Nature 349:796–799
60. Labow MA, Norton CR, Rumberger JM, Lombard-Gillooly KM, Shuster DJ, Hubbard J, Bertko R, Knaack PA, Terry RW, Harbison ML et al (1994) Characterization of E-selectin-deficient mice: demonstration of overlapping function of the endothelial selectins. Immunity 1:709–720
61. Gallatin WM, Weissman IL, Butcher EC (1983) A cell-surface molecule involved in organ-specific homing of lymphocytes. Nature 304:30–34
62. Lewinsohn DM, Bargatze RF, Butcher EC (1987) Leukocyte-endothelial cell recognition: evidence of a common molecular mechanism shared by neutrophils, lymphocytes, and other leukocytes. J Immunol 138:4313–4321
63. Ley K, Gaehtgens P, Fennie C, Singer MS, Lasky LA, Rosen SD (1991) Lectin-like cell adhesion molecule 1 mediates leukocyte rolling in mesenteric venules in vivo. Blood 77:2553–2555
64. Ley K, Tedder TF, Kansas GS (1993) L-selectin can mediate leukocyte rolling in untreated mesenteric venules in vivo independent of E- or P-selectin. Blood 82:1632–1638
65. Arbones ML, Ord DC, Ley K, Ratech H, Maynard-Curry C, Otten G et al (1994) Lymphocyte homing and leukocyte rolling and migration are impaired in L-selectin (CD62L) deficient mice. Immunity 1:247–260
66. Geng JG, Bevilacqua MP, Moore KL, McIntyre TM, Prescott SM, Kim JM et al (1990) Rapid neutrophil adhesion to activated endothelium mediated by GMP-140. Nature 343:757–760
67. Johnston GI, Cook RG, McEver RP (1989) Cloning of GMP-140, a granule membrane protein of platelets and endothelium: Sequence similarity to proteins involved in cell adhesion and inflammation. Cell 56:1033–1044
68. Palabrica T, Lobb R, Furie BC, Aronovitz M, Benjamin C, Hsu Y-M et al (1992) Leukocyte accumulation promoting fibrin deposition is mediated in vivo by P-selectin on adherent platelets. Nature 359:848–851
69. Lawrence MB, Springer TA (1991) Leukocytes roll on a selectin at physiologic flow rates: Distinction from and prerequisite for adhesion through integrins. Cell 65:859–873
70. Mayadas TN, Johnson RC, Rayburn H, Hynes RO, Wagner DD (1993) Leukocyte rolling and extravasation are severely compromised in P selectin-deficient mice. Cell 74:541–554
71. Bullard DC, Qin L, Lorenzo I, Quinlin WM, Doyle NA, Bosse R, Vestweber D et al (1995) P-selectin/ICAM-1 double mutant mice: acute emigration of neutrophils into the peritoneum is completely absent but is normal in pulmonary alveoli. J Clin Invest 95:1782–1788
72. Dustin ML, Rothlein R, Bhan AK, Dinarello CA, Springer TA (1986) Induction by IL-1 and interferon-gamma: tissue distribution, biochemistry, and function of a natural adherence molecule (ICAM-1). J Immunol 137:245–254
73. Simmons DL, Makgoba MW, Seed B (1988) ICAM, an adhesion ligand of LFA-1, is homologous to the neural cell adhesion molecule NCAM. Nature 331:624–627
74. De Fougerolles AR, Stacker SA, Schwarting R, Springer TA (1991) Characterization of ICAM-2 and evidence for a third counter-receptor for LFA-1. J Exp Med 174:253–267
75. Staunton DE, Dustin ML, Springer TA (1989) Functional cloning of ICAM-2, a cell adhesion ligand for LFA-1 homologous to ICAM-1. Nature 339:61–64
76. De Fougerolles AR, Springer TA (1992) Intercellular adhesion molecule-3, a third adhesion counter- receptor for lymphocyte function-associated molecule-1 on resting lymphocytes. J Exp Med 175:185–190
77. Fawcett J, Holness CL, Needham LA, Turley H, Gatter KC, Mason DY et al (1992) Molecular cloning of ICAM-3, a third ligand for LFA-1, constitutively expressed on resting leukocytes. Nature 360:481–484

78. Elices MJ, Osborn L, Takada Y, Crouse C, Luhowskyj S, Hemler ME et al (1990) VCAM-1 on activated endothelium interacts with the leukocyte integrin VLA-4 at a site distinct from the VLA-4/fibronectin binding site. Cell 60:577–584
79. Osborn L, Hession C, Tizard R, Vassallo C, Luhowskyj S, Chi-Rosso G et al (1989) Direct expression cloning of vascular cell adhesion molecule 1, a cytokine-induced endothelial protein that binds to lymphocytes. Cell 59:1203–1211
80. Cybulsky MI, Gimbrone MAJ (1991) Endothelial expression of a mononuclear leukocyte adhesion molecule during atherogenesis. Science 251:788–791
81. Muller WA, Weigl SA, Deng X, Phillips DM (1993) PECAM-1 is required for transendothelial migration of leukocytes. J Exp Med 178:449–460
82. Vaporciyan AA, DeLisser HM, Yan H-C, Mendiguren II, Thom SR, Jones ML et al (1993) Involvement of platelet-endothelial cell adhesion molecule-1 in neutrophil recruitment in vivo. Science 262:1580–1582
83. Berlin C, Berg EL, Briskin MJ, Andrew DP, Kilshaw PJ, Holzmann B et al (1993) α4β7 integrin mediates lymphocyte binding to the mucosal vascular addressin MAdCAM-1. Cell 74:185–195
84. Briskin MJ, McEvoy LM, Butcher EC (1993) MAdCAM-1 has homology to immunoglobulin and mucinlike adhesion receptors and to IgA1. Nature 363:461–464
85. Sako D, Chang X-J, Barone KM, Vachino G, White HM, Shaw G et al (1993) Expression cloning of a functional glycoprotein ligand for P-selectin. Cell 75:1179–1186
86. Hemmerich S, Bertozzi CR, Leffler H, Rosen SD (1994) Identification of the sulfated monosaccharides of GlyCAM-1, an endothelial-derived ligand for L-selectin. Biochemistry 33:4820–4829
87. Lasky LA, Singer MS, Dowbenko D, Imai Y, Henzel WJ, Grimley C et al (1992) An endothelial ligand for L-selectin is a novel mucin-like molecule. Cell 69:927–938
88. Baumhueter S, Singer MS, Henzel W, Hemmerich S, Renz M, Rosen SD et al (1993) Binding of L-selectin to the vascular sialomucin CD34. Science 262:436–438
89. Lenter M, Levinovitz A, Isenmann S, Vestweber D (1994) Monospecific and common glycoprotein ligands for E- and P-selectin in myeloid cells. J Cell Biol 125:471–481
90. Levinovitz A, Mühlhoff J, Isenmann S, Vestweber D (1993) Identification of a glycoprotein ligand for E-selectin on mouse myeloid cells. J Cell Biol 121:449–459
91. Brady HR, Spertini O, Jimenez W, Brenner BM, Marsden PA, Tedder TF (1992) Neutrophils, monocytes, and lymphocytes bind to cytokine-activated kidney glomerular endothelial cells through L-selectin (LAM-1) in vitro. J Immunol 149:2437–2444
92. Spertini O, Luscinskas FW, Kansas GS, Munro JM, Griffin JD, Gimbrone MAJ et al (1991) Leukocyte adhesion molecule-1 (LAM-1) interacts with an inducible endothelial cell ligand to support leukocyte adhesion. J Immunol 147:2565–2573
93. MacKay CR (1993) Homing of naive, memory and effector lymphocytes. Curr Opin Immunol 5:423–427
94. Parker CM, Cepek KL, Russell GJ, Shaw SK, Posnett DN, Schwarting R et al (1992) A family of β_7 integrins on human mucosal lymphocytes. Proc Natl Acad Sci USA 89:1924–1928
95. Berg EL, Goldstein LA, Jutila MA, Nakache M, Picker LJ, Streeter PR et al (1989) Homing receptors and vascular addressins: Cell adhesion molecules that direct lymphocyte traffic. Immunol Rev 108:5–18
96. Hamann A, Jablonski-Westrich D, Jonas P, Thiele HG (1991) Homing receptors reexamined: mouse LECAM-1 (MEL-14 antigen) is involved in lymphocyte migration into gut-associated lymphoid tissue. Eur J Immunol 21:2925–2929
97. Xu H, Gonzalo JA, St Pierre Y, Williams IR, Kupper TS, Cotran RS et al (1994) Leukocytosis and resistance to septic shock in intercellular adhesion molecule 1-deficient mice. J Exp Med 180:95–109
98. Steegmaier M, Levinovitz A, Isenmann S, Borges E, Lenter M, Kocher HP, Kleuser B, Vestveber D (1995) The E-selectin ligand ESL-1 is a variant of a receptor for fibroblast growth factor. Nature 373:615–620

Effects of *N*-Acetyl-Cysteine on Splanchnic Circulation

A. G. N. Agustí, J. Ibañez, and B. Togores

Introduction

In stressful circumstances (hermorrhage, sepsis), several endogenous vasoconstrictors (cathecholamines, angiostensin II, leukotriene D_4, thromboxane A_2) are released into the circulation [1]. These vasonconstrictors affect the splanchnic circulation more consistently than other tissue beds [2–4]. Their net effect is to redistribute blood away from the splanchnic organs to preserve blood flow to the so-called vital organs (brain and heart) [1]. Splanchnic ischemia is frequently observed in experimental models of sepsis [4] and in critically ill subjects [1]. It has been suggested that splanchnic ischemia can be an important contributor to the development of multiple organ failure (MOF), particularly during sepsis [1]. In critically ill patients the detection and treatment of splanchnic ischemia is clinically relevant because it seems to have prognostic implications [5]. This chapter reviews the available methods to detect splanchnic ischemia clinically and the effects of *N*-acetyl-cisteine (NAC) on whole body tissue oxygenation and splanchnic oxygenation, both in experimental animals and in critically ill patients.

Clinical Monitoring of Splanchnic Circulation

Ischemia of the gastric mucosa may be an early manifestation of impaired tissue perfusion in critically ill patients [3]. Measurement of gastric mucosal pH (pHi) by tonometry allows appropriate monitoring of gastric oxygenation in these patients [1, 5–7]. Importantly, measurement of pHi can detect changes in tissue oxygenation well before any other index of whole body oxygenation (e.g., oxygen uptake, oxygen delivery, extraction ratio) had changed significantly [6, 8]. The measurement of pHi by gastric tonometry employs a standard vented nasogastric sump that has a distal silicone balloon [1, 9]. The gastric tonometer is inserted into the stomach as a standard nasogastric tube. Once in place, the balloon lies close to the gastric mucosa. Because it is permeable to CO_2, it functions as a tonometer when filled with saline [1, 9, 10]. After an equilibration period the PCO_2 of the saline reflects the PCO_2 of the gastric mucosa [1, 9, 10]. The saline PCO_2 (and the bicarbonate of a simultaneously obtained arterial blood sample) is used in the

Henderson-Hasselbalch equation to calculate the pHi [1, 9]. Several studies have stressed the clinical utility of pHi monitoring in critically ill patients [5–7].

Effects of *N*-Acetyl-Cysteine on Tissue Oxygenation

In healthy individuals the metabolic activity of cells is largely independent of the supply of oxygen by blood [11]. However, during sepsis several authors have shown a defect in tissue O_2 extraction ability that renders O_2 uptake (VO_2) pathologically dependent on O_2 delivery (QO_2) [3, 12, 13]. Microvascular obstruction by activated inflammatory cells, release of vasoactive and toxic substances and endothelial cell damage can all contribute to this O_2 extraction defect of sepsis [14–19]. Oxygen free radicals may also be important in this scenario [14, 16] because they seem to cause endothelium dysfunction and myocardial depression in sepsis [20]. NAC has important antioxidant properties [20]. In addition, NAC blocks the effects of tumor necrosis factor, enhances nitric oxide release by the endothelium, and decreases neutrophil-aggregating activity [20]. Thus, in theory NAC might be useful in patients with sepsis and abnormal tissue O_2 extraction.

Zhang et al. have recently explored this hypothesis in septic dogs, where systemic blood flow was progressively reduced by cardiac tamponade [20]. In this study NAC was administered before endotoxin was given. The main findings in this investigation were that, compared to septic control animals, those receiving NAC showed more systemic (and pulmonary) vasodilation, higher cardiac output, and higher VO_2 and QO_2 at any level of intrapericardial pressure [20]. The critical extraction ratio and the slope of the VO_2/QO_2 relationship were higher in NAC-treated animals than in controls. Finally, NAC decreased lactate and tumor necrosis factor levels and also increased glutathione peroxidase activity [20]. To summarize, this elegant study shows that the administration of NAC before endotoxin improved O_2 extraction by tissues and myocardial function in septic dogs [20]. However, as the authors themselves conclude, an important clinical question remains whether NAC can exert these protective effects once the septic response has already developed [20]. Another important question, implicit in the paper, is whether these experimental findings are applicable to critically ill humans.

In humans intravenous NAC is the treatment of choice in fulminant hepatic failure associated with acetaminophen overdose [21]. Harrison et al. have reported the effects of NAC on tissue oxygenation in 12 patients with aceptaminophen-induced fulminant hepatic failure [21]. Similarly to the above experimental findings [20], these authors observed that NAC increases cardiac output, lowers the systemic vascular resistance, augments QO_2 and VO_2, and, importantly, enhances O_2 extraction by tissues [21]. The underlying mechanisms were not revealed by the study of Harrison et al., but the authors suggested that NAC may enhance the activity of the endothelium-derived relaxing factor (nitric oxide) at the microcirculatory level [21]. In theory the higher VO_2 observed after NAC in this study suggested improved cell bioenergetics and was explained by two different mechanisms: (a) increased *convective* O_2 transport (because of the higher cardiac output), and (b) increased *diffusive* O_2 transport (as shown by the higher O_2 extraction ratio after NAC) [21]. This latter mechanism is of great pathophysiological relevance [22] and suggests that NAC possibly facilitates a better match between O_2 demand and O_2 supply in these patients. It should be noted, however, that these patients with liver failure had associated a severe hyperdynamic cardiocirculatory state and therefore a very low O_2 extraction ratio [21]. This factor may have important pathophysiological consequences (see below).

Our group has recently reported the preliminary results of a study aimed at evaluating the short-term effects of NAC upon whole body and splanchnic oxygenation in patients with severe MOF and evidence of splanchnic ischemia (as assessed by gastric tonometry) [23]. Similarly to the results reported by Harrison et al. [21], in patients with MOF NAC also increases cardiac output, vasodilates pulmonary and systemic circulation, increases oxygen delivery to tissues, and increases oxygen uptake (measured directly using expired gases) [23]. These observations are also similar to those reported in septic dogs [20]. However, contrary to previous studies, in patients with MOF NAC did not enhance the extraction of O_2 by tissues [23]. An important difference between our study and that reported by Harrison et al. in patients with liver failure [21], is the absolute value of the O_2 extraction fraction – in our patients it was normal and much higher than that reported in patients with liver failure. The higher hyperdynamic cardiocirculatory state associated to liver failure may have influenced significantly the behavior of the microcirculation [24] and thus contributed to the different effect of NAC upon tissue O_2 extraction in these two different clinical situations.

Conclusion: Effects of *N*-Acetyl-Cysteine on the Splanchnic Circulation

There are only preliminary reports of the effects of NAC on the splanchnic circulation. Zhang et al. studied the effects of NAC on femoral, renal, and mesenteric blood flow when cardiac output was reduced progressively by cardiac tamponade in septic dogs [25]. They showed that, compared to septic controls, NAC improved myocardial function and mesenteric, renal, and femoral blood flow [25]. We evaluated the effects of NAC on the splanchnic circulation in patients with MOF by monitoring the changes in pHi after a short-term infusion of the drug [23]. In these patients we observed that pHi was not improved by NAC. In fact, pHi decreased sligtly but signicantly after NAC [23]. Although these are preliminary observations, we speculate that in critically ill patients NAC may interfere with the mechanisms that control blood flow distribution among organs [26], vasodilating other territories more than splanchnic circulation and stealing blood from the gastric circulation. This hypothesis should be confirmed in future studies.

Acknowledgments. The authors thank Dr. A. Maimo (Servei Pneumologia), and Drs. J.M. Raurich, P. Marsé, R. Jordá y J. Bergadá (Servei Cures intensives) of Hospital Univ. Son Dureta (Palma Mallorca) for their cooperation in this study. This work been partially supported by SEPAR, FISss (93/0851), ABEMAR, and Zambón SA.

References

1. Fiddian-Green RG (1989) Studies in splachnic ischemia and multiple organ failure. In: Bulkley MA, Fiddian-Green RG, Haglund U (eds) Splachnic ischemia and multiple organ failure. Mosby, St Louis, chap 32, pp 349–363
2. Nelson DP, King CE, Dodd SL, Schumacker PT, Cain SM (1987) Systemic and intestinal limits of O2 extraction in the dog. J Appl Physiol 63:387–394
3. Nelson DP, Samsel RW, Wood LDH, Schumacker PT (1988) Pathological supply dependence of systemic and intestinal O2 uptake during endotoxemia. J Appl Physiol 64:2410–2419
4. Vallet B, Lund N, Curtis SE, Kelly D, Cain SM (1994) Gut and muscle tissue PO_2 in endotoxemic dogs during shock and resuscitation. J Appl Physiol 76:793–800
5. Gutierrez G, Palizas F, Doglio G et al (1992) Gastric intramucosal pH as a therapeutic index of tissue oxygenation in critically ill patients. Lancet 339:195–199
6. Maynard N, Bihari D, Beale R et al (1993) Assessment of splanchnic oxygenation by gastric tonometry in patients with acute circulatory failure. JAMA 270:1203–1210
7. Schlichtig R, Stremple JF (1994) Measuring splanchnic oxygenation by gastric tonometry. JAMA 271:983–984
8. Ruokonen E, Takala J, Kari A, Saxén H, Mertsola J, Hansen EJ (1993) Regional blood flow and oxygen transport in septic shock. Crit Care Med 21:1296–1303
9. Clark CH, Gutierrez G (1992) Gastric intramucosal pH: a noninvasive method for the indirect measurement of tissue oxygenation. Am J Crit Care 2:53–60
10. Fiddian-Green RG, Pittenger G, Whitehouse WM (1982) Back-diffusion of CO2 and its influence on the intramural pH in gastric mucosa. J Surg 33:39–48
11. Schumacker PT, Samsel RW (1989) Oxygen delivery and uptake by peripheral tissues: physiology and pathophysiology. Crit Care Clin 5:255–269
12. Nelson DP, Beyer C, Samsel RW, Wood LDH, Schumacker PT (1987) Pathological supply dependence of O2 uptake during bacteremia in dogs. J Appl Physiol 63:1487–1492
13. Samsel RW, Nelson DP, Sanders WM, Wood LDH, Schumacker PT (1988) Effect of endotoxin on systemic and skeletal muscle O2 extraction. J Appl Physiol 65:1377–1382
14. Leff JA, Parsons PE, Day CE et al (1993) Serum antioxidants as predictors of adult respiratory distress syndrome in patients with sepsis. Lancet 341:777–780
15. Lübbe AS, Garrison RN, Cryer HM, Alsip NL, Harris PD (1992) EDRF as a possible mediator of sepsis-induced arteriolar dilation in skeletal muscle. Am J Physiol (Heart Circ Physiol) 262:H880–H887
16. Moncada S, Higgs A (1993) The L-arginine-nitric oxide pathway. N Engl J Med 329:2002–2012
17. Nahum A, Chamberlin W, Sznajder JI (1991) Differential activation of mixed venous and arterial neutrophils in patints with sepsis syndrome and acute lung injury. Am Rev Respir Dis 143:1083–1087

18. Umans JG, Wylam ME, Samsel RW, Edwards J, Schumacker PT (1993) Effects of endotoxin in vivo on endothelial and smooth-muscle function in rabbit and rat aorta. Am Rev Respir Dis 148:1638–1645
19. Wylam ME, Samsel RW, Umans JG, Mitchell RW, Leff AR, Schumacker PT (1990) Endotoxin impairs endothelium-dependent relaxation of canine arteries in vitro. Am Rev Respir Dis 142:1263–1267
20. Zhang H, Spapen H, Nguyen DN, Benlabed M, Buurman WA, Vincent J-L (1994) Protective effects of N-acetyl-L-cysteine in endotoxemia. Am J Physiol (Heart Circ Physiol) 266:H1746–H1754
21. Harrison PM, Wendon JA, Gimson AES, Alexander GJM, Williams R (1991) Improvement by acetylcysteine of hemodynamics and oxygen transport in fulminant hepatic failure. N Engl J Med 324:1852–1857
22. Smithies M, Bihari DJ (1993) Delivery dependent oxygen consumption: asking the wrong questions and not getting any answer. Crit Care Med 21:1622–1626
23. Maimó A, Raurich JM, Togores B et al (1994) Effects of N-acetyl-cysteine (NAC) on tissue oxygenation in patients with multiple organ failure (MOF). Am J Respir Crit Care Med 149:A416
24. Agustí AG, Roca J, Bosch J, Rodriguez-Roisín R (1990) The lung in patients with cirrhosis. J Hepatol 10:251–257
25. Zhang H, Spapen H, Nguyen DN, Rogiers P, Van der Linden P, Vincent J (1994) Effects of N-Acetylcysteine on regional blood flow during endotoxic shock. Am J Respir Crit Care Med 149:A416
26. Schlichtig R, Kramer DJ, Pinsky MR (1991) Flow redistribution during progressive hemorrhage is a determinant of critical O2 delivery. J Appl Physiol 70:169–178

During the editorial process, the following reference related to the topic of this chapter has been published:
Spyes CD, Reinhart K, Witt I (1994) Influence of acetyl cysteine on indirect indicators of tissue oxygenation in septic shock patients: results from a prospective, randomized, double-blind study. Crit Care Med 22:1738–1746

Hepatic Acute-Phase Response

D. Thébert, A. Mignon, and J.-F. A. Dhainaut

Introduction

The acute-phase response (APR) refers to a large number of systemic and metabolic changes secondary to local inflammation following a wide variety of noxious stimuli. The principal stimuli that give rise to the APR are infection, surgical and other trauma, neoplasm, burn injury, tissue infarction, various inflammatory states, and childbirth. The APR is thought to play a major adaptive and defense role [1]. Because of its immunological and metabolic functions, the liver plays a crucial role in APR. This chapter reviews the early events following the initial trauma, the major soluble mediators, the principal systemic changes, and the liver APR and its regulation.

Inflammation Early Events

The classic manifestations of local acute inflammation – redness and swelling with heat and pain – have been recognized for nearly 2000 years. These clinical signs may be due to changes in either cellular and humoral components. After the initial trauma, complex interactions take place in the damaged tissue between the formed elements of the blood (platelets, monocytes/macrophages, polymorphonuclears), capillary endothelium, and stromal cells. These interactions depend on soluble mediators released by the various protagonists, including eicosanoids, cytokines, platelet-activating factor, and thromboxane A_2. This contributes to the host defense. The monocyte-macrophage cell population is upregulated by the cytokine [interleukin (IL)-1, tumor necrosis factor (TNF)] production that induces gene expression of important adhesion molecules, including intercellular adhesion molecules in the endothelial cells. These molecules interact specifically with circulating leukocytes to slow their rate of flow, initiate transendothelial passage and allow subsequent migration to the tissue [2], where they produce large quantities of superoxide and hydroxyl radicals (via a state of enhanced potential respiratory burst activation), and proteases. In another way, fibroblasts, activated by TNF, and IL-1, can also produce collagenase, nitric oxide (NO) and superoxide anions. The release of such radicals and proteases promotes bacterial and damaged host cell destruction. The net result of the enhanced capillary permeability, mediated by IL-1 via platelet-activating factor and prostaglandin secretions, is to allow the increased circulation of mediators which act to recruit leukocytes via chemotatic factors (IL-8, monocyte chemoattractant protein), and activate systemic acute-phase reaction.

Principal Mediators of the Acute-Phase Response

The three major mediators involved in the APR are TNF, IL-1, and IL-6 (properties reviewed in Table 1).

TNF and IL-1

The term TNF was coined in 1975 when Carswell et al. reported that infusion lipopolysaccharide (LPS)-induced shock serum into mice primed by injection of bacillus Calmette-Guérin induces hemorrhagic necrosis of solid tumor [3]. TNFα (reviewed in [4]) is a trimeric molecule of 17-kDa subunits of 154–157 amino acid residues. It is synthesized in LPS- or interferon (IFN)-activated monocytes as a prehormone and undergoes cleavage at several distinct sites during posttranslational processing to yield to a mature protein. Several investigators have described a membrane-associated TNF that may play a role in cell-to-cell or other short range interactions. TNFα is encoded by a gene within the class III region of the major histocompatibility complex on the short arm of chromosome 6. Its expression is highly regulated. Its transcription and the half-life of its mRNA increase in response to LPS. Glucocorti-

Table 1. Comparison of IL-1, TNF, and IL-6 (from [54])

Biological property	IL-1	NF	IL-6
Endogenous pyrogen fever	+	+	+
Slow-wave sleep	+	+	−
Hepatic acute-phase proteins	+	+	+
T-cell activation	+	+	+
B-cell activation	+	+	+
B-cell Ig synthesis	−	−	+
Fibroplast proliferation	+	+	−
Stem-cell activation (hemopoietin-1)	+	−	+
Nonspecific resistance to infection	+	+	+
Radioprotection	+	+	−
Cyclo-oxygenase, PLA_2, gene expression	+	+	−
Synovial cell activation	+	+	−
Endothelial cell activation	+	+	−
Shock syndrome	+	+	−
Induction of IL-1, TNF, and IL-8	+	+	−
Induction of IL-6	+	+	−

costeroids suppress TNF gene expression by both transcriptional and post-transcriptional mechanisms (TNF mRNA are rapidly degraded because of a high sensibility to ribonucleases, which depends on repeating octameric element in the 3' untranslated region.). TNF has been shown to mediate many of the systemic manifestations of gram-negative sepsis, including fever and hypotension. It has many other biological effects including autocrine or paracrine activation of macrophages and induction of the release of other cytokines – IL-1 and IL-6 – and growth factors – platelet-derived growth factor (PDGF) and granulocyte-macrophage colony-stimulating factor (GM-CSF; reviewed in [5]). It has direct effects on hepatic acute-phase gene expression that are similar to those of IL-1β, but involves distinct cell surface receptor and signal transduction pathways.

IL-1 (reviewed in [6]) is synthesized by a wide variety of cell types: macrophages, fibroblasts, keratinocytes, renal mesangial cells, vascular endothelial and smooth muscle cells, B lymphocytes, natural killer cells, and some T lymphocyte cell lines. Two distinct molecules (IL-1α and IL-1β) are encoded by two distinct genes on chromosome 2. IL-1 is the predominant form in humans, with a molecular weight of 17.4 kDa; this may be due to its promoter, which is 10- to 50-fold more efficient than that of the α gene [7]. IL-1α and -1β share relatively little amino acid homology (approximately 26% in humans). Both forms of IL-1 are synthesized as 31-kDa precursor molecules and undergo proteolytic processing to generate the 17.4-kDa mature protein. Two distinct receptors have been identified [8], representing two different gene products. One is found on T cell, fibroblasts, and epithelial cells, the other one on B-lymphocytes, macrophages, and bone marrow cells. The IL-1 receptor of hepatocytes has binding properties that are closer to that of B-lymphocytes and macrophages. Each receptor can bind the both forms of IL-1. IL-1 has many biological activities. It has been shown to elicit fever, increase sensation of pain, slow-wave sleep, induce T-lymphocytes and synovial fibroblasts proliferation, secretion of prostaglandins and metalloprotease in synovial fibroblasts, expression of leucocyte adherence receptors in endothelial cells, stimulate its own production, as well as IL-2, IL-3, IL-6, and IFN. IL-1 activates several acute-phase liver gene transcription.

Interleukin-6

IL-6 comprises a group of polypeptides derived from a single gene on chromosome 7 which differ in molecular weight, glycosylation, serine phosphorylation, and sulfatation. A 45-kDa polypeptide is the major species observed in human plasma by western blot analysis. IL-6 can be detected in normal human plasma, but its secretion increases in the presence of endotoxin, IL-1, TNF, PDGF, IFN-γ, phorbol ester, and agents that increase cAMP [9, 10]. Glucocorticosteroids inhibit the regulation of IL-6 gene expression by TNF, IL-1, and LPS in fibroblasts. IL-6 is synthesized by mononuclear phagocytes, T and B lymphocytes, keratinocytes, endothelial cells, and stromal cells (mesangial renal cells, fibroblasts, endometrial stromal cells). As IL-1, IL-6 has pleiotropic biological activities. It is involved in B lymphocytes and bone marrow stem cells maturation, T lymphocytes proliferation, and terminal differentiation of monocytes and sympathetic neurones. It induces IL-2 receptor and IL-2 gene expression in lymphocytes. The regulation of hepatic acute-phase gene expression is one of its important functions.

Systemic Effects of Inflammatory Mediators: The Acute-Phase Response

The APR reviewed in [11] takes place within hours following the inflammatory stimuli. Characteristics of the systemic response include fever, alterations in vascular permeability, leukocytosis, increased plasma copper and decreased plasma iron and zinc, increased hormone level (e.g., glucocorticosteroids), gluconeogenesis, and increased plasma glycoprotein levels [1]. Fever is one of the physiological responses to inflammation. The putative role of IL-1 has been recognized for years [6]. It is now thought that IL-1, as well as TNF and IL-6, are involved in the febrile response. These cytokines trigger fever by increasing prostaglandin E_2 synthesis in the anterior hypothalamic region. Recent studies have shown IL-1α, IL-1β, and TNF to be somnogenic on injection into the lateral cerebral ventricle of rabbits. IL-1 has been shown to increase slow-wave sleep.

Capillary endothelial cell is a usual target for TNF and IL-1. Polymorphonuclear cells which synthetize hydroxyl radicals, platelet-activating factor, prostaglandins, and C5a complement component are involved in this lesional process and can lead to hemorrhagic necrosis in normal tissue [12–14]. Many transport proteins of mineral and trace elements are acute-phase proteins (APPs). Changes in their synthesis explain changes in the plasma levels of iron, copper, and zinc. Metallothioneine gene expression is increased following IL-1 and TNFα and produces a transient depression of zinc in plasma and a concomitant uptake in the liver. In human HepG2 hepatoma cell, IL-1 induces metallothioneine synthesis. Hepatic induction of ceruloplasmin synthesis by IL-1 and IL-6 leads to a rise in the serum copper level. Transferrin synthesis is downregulated in the liver by IL-1, IL-6, and TNF. IL-1 and IL-6 can act on the adrenal pituitary axis, via the production of corticotropin-releasing factor by hypothalamus, to generate adrenocorticotropic hormone and cortisol [15]. Glucocorticoids play a permissive and synergistic role in the induction of hepatic APP expression, and exert a feedback inhibition on cytokine production in inflammatory cells.

TNF has been shown to stimulate bone resorption and inhibit bone formation. Inflammation mediators can also interact with coagulation processes.In vivo TNF induces the early appearance of activated factor X for a short period, then a more progressive and more prolonged enhancement of the plasmatic level of one fragment of prothrombin [16]. TNF and IL-1 can affect both in vivo and in vitro the fibrinolysis processes by inducing the production of an inhibitor of tissue-type plasminogene activator by endothelial cells. This could explain the role of cytokines in microvascular thrombosis.

Important metabolic changes also occur in which the liver plays a pivotal role. In vivo IL-1 and TNFα induce an increase in protein catabolism in the muscle and a simultaneous rise of amino acid uptake by hepatocytes leading to an amino acid shift from the muscle to the liver [17, 18]. In primary culture, however, only IL-6 – and not IL-1 or TNF – stimulates amino acid uptake into hepatocytes [19]; IL-1 increases the degradation of amino acids and the formation of biogenic amines by inducing histidine decarboxylase and ornithine decarboxylase activities in the liver. TNF, given alone, however, is less potent but further enhances the IL-1β induced histidine and ornithine decarboxylase activities. Inflammatory states are often associated with an increase in the plasmatic levels of free fatty acids and with hyperlipemia. Hepatic fatty acid synthesis in vivo is found to be increased by TNFα and IL-1; cholesterol synthesis is increased after TNFα and IL-1 exposition. Lipoprotein lipase gene expression in enhanced by TNFα.

Cytokines modulate carbohydrate metabolism. Glucose release is stimulated by IL-6 in primary cultures of hepatocytes. IL-1 has direct cytotoxicity on β cells of islets of Langerhans. It inhibits glucose modulated insulin and glucagon secretion but can stimulate insulin secretion at low concentration. Several cytokines such as IL-1α, IL-1β, TNFα, and TNFβ have been found to inhibit hepatic cytochrome P-450 and related drug metabolizing *in vivo* [20].

Hepatic Acute-Phase Reaction

The APR is defined by the qualitative and quantitative changes in the expression of plasma protein following tissue injury [1]. These proteins can be divided into two classes: (a) the positive APPs whose plasma concentration increase during acute-phase reaction, for example, C-reactive protein (CRP), α_1-acid glycoprotein, haptoglobin, serum amyloid A (SAA) and P, hemopexin and fibrinogen, and the major antiproteases (α_1-antitrypsin, α_1-antichymotrypsin, thiostatin, and α_2-macroglobulin in rats); (b) the negative acute-phase reactants whose levels decrease, for example, albumin and transferrin.

Tissue injury leading to activation of liver metabolism has suggested the presence of soluble mediators [21]. In fact, several cytokines and hormones have been shown to be involved in the regulation of hepatic APR. According to the pattern of induction by cytokines, APP genes have been divided into two major categories in hepatic cell cultures [22]: (a) type-1, regulated by IL-1 type cytokines (e.g., IL-1α, IL-1β, TNFα, TNFβ), includes α_1-acid glycoprotein, SAA, CRP, complement component C3 and haptoglobin; (b) type-2, regulated by IL-6 type cytokines [e.g., IL-6, IL-11, leukemia inhibitory factor (LIF), oncostatin M, and cilliary neurotrophic factor], including, in most species, fibrinogen, haptoglobin, hemopexin, α_2-macroglobulin, thiostatin, α_1-antichymotrypsin, α_1-antitrypsin.

IL-11, LIF, oncostatin M, and cilliary neurotropic factor differ structurally from IL-6 but are involved in the stimulation of the same spectrum of APP genes. Moreover, LIF and IL-6 induce the same pattern of early cytoplasmic tyrosine phosphorylation, supporting the possibility that a common signal transduction pathway could be shared by all these mediators [23]. Type-1 APP genes are synergistically enhanced by IL-6 type cytokines [22]. IL-1 type cytokines do not stimulate type-2 APP genes and do not enhance the effect of IL-6 type cytokines on type-2 APP genes. Glucocorticosteroids, insulin, and growth factors are involved as modulators of cytokine action. Glucocorticoids are able to stimulate expression of some APPs, especially in the rat; however, their principal action is to exert a permissive and synergistic action on the effect of IL-1 and IL-6 types of cytokines (the regulation of hepatic APP production is summarized in Table 2).

Table 2. Regulation of hepatic-APP production (from [11])

Effectors	APP response
IL-1-type cytokines (IL-1α, IL-1β, TNFα, TNFβ)	Stimulation of type 1 APP C-reactive protein (human) Serum amyloid A α_1-Acid glycoprotein Complement C3 Haptoglobin (rat) Hemopexin (rat) Inhibition of type 2 APP
IL-6-type cytokines (IL-6, IL-11, LIF, OSM, CNTF)	Stimulation of most APP Primary activator of type 2 APP α-, β-, and γ-Fibrinogen α_2-Macroglobulin (rat) Thiostatin (rat) α_1-Antitrypsin α_1-Antichymotrypsin (or contrapsin in rodents) Haptoglobin (human) Hemopexin (human) Synergism with IL-1-type cytokines on type 1 APP
Glucocorticoids	Minor stimulation of most APPs Stimulation of rat α_1-acid glycoprotein Strong synergistic enhancement of cytokine effects on most APP
Insulin	Reduction of IL-1 and IL-6 types of cytokine effects on most APP
HGF, FGF	Minor reduction in IL-1 and IL-6 types of cytokine effects on most APP Enhancement of IL-1 type of cytokine effects on rat α_1-acid glycoprotein and complement C3
Hormones acting via cyclic nucleotides or calcium mobilization	No detectable regulatory effect on APP

APP, Acute-phase plasma protein; IL, interleukin; TNF, tumor necrosis factor; LIF, leukemia inhibitory factor; OSM, oncostatin M; CNTF, ciliary neurotropic factor; HGF, hepatic growth factor; FGF, fibroblast growth factor.

CRP is the first APP discovered. In 1930 Tillet and Francis isolated a protein in sera of patients with pneumococcal pneumonia which precipitated with the C cell wall polysaccharide of the pneumococcus, the CRP [24]. CRP is one of the two pentraxins, with the serum amyloid P component, isolated in human plasma. It is a discoid cyclic molecule composed of five nonconvalently associated subunits [25]. Each subunit is a single chain of 206 amino acid with one interchain disulfide bond. Its gene takes place on long arm of chromosome 1. CRP is present at 0.1 mg/dl concentration or less in normal adult sera; its level increases 100-fold upon tissue injury or inflammation [1]. Its function is unknown, but highly conserved [26] and the lack of any known protein polymorphism suggest an important biological function. Its has been found to be involved in complement binding, chemoattraction, and opsonization [27]. It also binds to chromatin [28] and phosphorylcholine moieties.

SAA protein was discovered in human sera using antisera raised to a protein purified from secondary amyloidosis fibrils (AA protein). SAA is a 12-kDa polypeptide that circulates in association with a subclass of high-density lipoproteins [29]. Its function is unknown, but it may be involved in the pathogenesis of amyloidosis [30, 31]. There are three isoforms of SAA, three different mRNAs, and probably three genes. Several physiological differences exist between the three isoforms: SAA_1 and SAA_2 are high-density lipoproteins apoproteins; SAA_3 has not been identified in serum. Native 8-kDa amyloid A-protein is the major isoform of insoluble fiber of inflammation associated amyloid deposit and is thought to be derived from SAA_2 by proteolytic processing [32]. The SAA plasmatic level increases during APR more than 100- to 1000-fold.

Fibrinogen is a 340-kDa hexamer, composed of three pairs of disulfide-linked subunits ($[Aa, Bb, \gamma]_2$), and encoded by three different genes on human chromosome 4 [33]. Two different mRNA for the two different subunits arise by alternative splicing [34]. Its level increases 2.5-fold after stimulus for a period of several weeks. It plays a pivotal role in the coagulation cascade.

Human plasma contain three major antiproteases α_1-antitrypsin (also known as α_1-proteinase inhibitor, API), α_1-antichymotrypsin and α_2-macroglobulin. API is a single-chain glycoprotein of 55 kDa and belongs to serine proteinase inhibitor (serpin) supergene family. Its gene takes place on chromosome 14. API binds rapidly to neutrophil elastase. Its plasmatic level increases three- to fourfold after stimulus. α_1-antichymotrypsin is a single-chain glycoprotein of 58 kDa encoded by a gene on chromosome 14 adjacent to the API gene. Its physiological function is unknown, but it belongs to the serpin supergene family and rapidly inactivate cathepsin G. Its plasmatic level increases rapidly during APR, reaching a level four- to fivefold basal.

Complement factor B is an approximately 93-kDa single-chain glycoprotein that belongs to the alternative-complement activating pathway C3 convertase. It is encoded by a single gene in the class III region of the major histocompatibility complex and secreted as a zymogen cleaved on activation by factor D. Its plasmatic level increases during inflammation. C3 complement component is known as a linchpin of the complement system because it is a substrate of both activation pathways. It is the most abundant complement component in the plasma (approx. 1 g/l), its level increases in APR. C3 is synthesized as a pro-C3. Pro-C3 is a two chains (α, β) disulfide-linked glycoprotein of 185 kDa, a posttranslational cleavage generates C3. C3 is hydrolyzed spontaneously in C3i or by C3 convertase in C3a and C3b. C3 has lectin properties and can bind to carbohydrate moieties of the *Staphylococcus epidermidis*, *Saccharomyces cerevisiae*, or *Histoplasma capsulatum* membranes. C3b is involved in C5-convertase production of both classical and alternative pathways. C3b and its cleavage products are important ligands in the interaction of cell surface receptors with other cell, micro-organism and molecules to which it has bound [35, 36]. C3a has chimiotactic properties and activates mast cell and basophil cell degranulation.

Human α_2-macroglobulin is a large Zn^{2+} binding glycoprotein of 718 kDa. Both humans and rat species express a constitutive α_2-macroglobulin in plasma, but during inflammation an inducible α_2-macroglobulin is also synthesized. In humans, in addition to the constitutive tetrameric α_2-macroglobulin, an inducible dimeric form is synthesized during severe inflammatory reaction and cancer. This form has been called pregnancy zone protein, because of its increase during pregnancy. In rat the basal level increases 100- to 1000-fold during APR [37, 38]. The exact function of α_2-macroglobulin is unknown. It has been found to bind to a large spectrum of mammalian and bacterial proteinases, but its antiproteinase function is of uncertain physiological importance because most proteinases interact preferentially with their specifiic inhibitors. These proteinases are internalized in a molecular trap, where they retain activity. They are sterically inaccessible to large protein substrates or antiproteinase, but they can cleave small peptides such as cytokines [39]. This internalization leads to conformational changes that expose domains that are recognized by receptors. These changes can also be produced by little amines such as methylamine. The proteinase-α_2-macroglobulin complex is the activated form of α_2-macroglobulin; after binding its receptor, it is rapidly endocytosed and removed from extracellular fluids [40]. This receptor has an identical amino acid sequence as the low-density lipoprotein receptor related protein [41, 42] and is expressed in various cell types such as hepatocytes, macrophages, and fibroblasts. Cytokines can bind to native form of α_2-macroglobulin as IL-6 [43] or to activated forms as IL-1β and TNFα [44, 45] and remain even partially active in various cell culture bioassays. α_2-Macroglobulin may influence cytokine activities in several ways; for example, through the binding of cytokines it protects them from glomerular filtration or proteinase inactivation, but, in the case of IL-1β or TNFα the complexes may be rapidly removed from the circulation by the cell which expresses α_2-macroglobulin receptor. Although in vitro studies have demonstrated that α_2-macroglobulin can bind IL-1β, TNFα, and IL-6 [43–45], endogenously generated complexes have not been identified in vivo.

Albumin is the most abundant plasmatic protein and constitutes more than 65% of the total plasma protein content in a normal adult human. It is a single-chain protein of 586 amino acids without carbohydrate moieties (reviewed in [46]). Albumin is encoded by a gene localized on chromosome 4 and linked to the genes of vitamin D binding protein and α-fetoprotein, which is the predominant plasma protein during fetal life. Albumin is synthesized as a preprotein of 610 amino acid; mature albumin is secreted after posttranslational cleavage. Albumin is involved in the maintenance of the oncotic pressure and in the transport of a wide variety of molecules. The albumin level decreases dramatically during APR. Its expression is also regulated by hormones such as thyroid hormone, glucorticosteroids [46], and insulin, which has been found to reduce 57-fold its mRNA levels *in vitro* by transcriptional and posttranscriptional mechanisms.

It is interesting to note that when the liver is the predominant site of synthesis, some APP can be produced in other cells in response to the same stimulating factor. Thus, α_1-antitrypsin can be synthetized by monocytes in vitro, and its expression is profoundly influenced by cytokines and other products generated during inflammation. These proteins may be involved in local host response to aggression to prevent ongoing tissue damage.

Hepatic Acute-Phase Regulation

The expression of most AAP genes is regulated at the mRNA level [47]. The intracellular events following the binding of a cytokine to its ligand can be divided in two classes: (a) immediate early events which are detected within few minutes and take place in the cytoplasm as a direct consequence of the receptor assembly and activation, and (b) early events which locate to the nucleus and implicate transcription factors in the regulation of genes. Transcription factors are sequence specific binding protein that initiate the formation of a transcription initiation complex on the promoter. These discrete regulatory DNA sequences, called *cis*-acting elements, are highly conserved, are located upstream from the gene promoter, and have been found to have a pivotal role in conferring extracellular stimuli into an enhanced rate of transcriptional initiation.

The receptors of the major mediators of APR have been identified. Receptor for IL-6 is composed of two different subunits with distinct functions which associate in the presence of the ligand. The receptor assembly is a multistep phenomenon: IL-6 binds a 80-kDa low-affinity IL-6 binding protein (IL-6Rα); then the complex induces the homodimerization of gp130 through a disulfide bond. IL-6Rα and gp130 belong to the hematopoietin supergene family. IL-6Rα has a very short intracellular domain and does not interact with the signal transmission. The gp130 intracellular domain is essential for signal transmission. The other members of the IL-6 related family share gp130; their receptors are composites formed by gp130 and a specific receptor, except for oncostatin M, which binds to gp130 with low affinity and induces its heterodimerization with LIF binding protein (LIF low-affinity receptor). Both gp130 homodimer and gp130-LIF binding protein heterodimer induce an activation of a receptor coupled tyrosine kinase. TNF receptor seems to be formed by several heteromeric subunits. The binding subunit is approximately 75 kDa, but the entire receptor is thought to be 300 kDa. Both TNFα and -β bind this receptor. AP-1, and NF-κB transactivating factors are thought to be involved in TNF signal transduction. IL-1 receptor is a single membrane-spanning segment, and three immunoglobulinlike domains form the extracellular ligand-binding region. It can bind both forms of IL-1 despite their poor homology (26% of primary structural homology). Little is known about signal transduction pathway. A serine/threonine protein kinase might be activated, and the IL-1 receptor itself is phophorylated after the binding of its ligand.

Several transcription factors have been shown to enhance APP transcription: NF-κB for type 1 genes, C/EBP and glucorticoids receptors for both type 1 and type 2, and IL-6REBP/acute-phase response factor (APRF) for type 2 APP genes. AP-1 is formed by the homo- or heterodimerization of two *jun* and *fos* family members and has an increased binding and transcriptional activity after IL-1, IL-6, or TNFα treatment. IL-1 causes a rapid but transient stimulation of c-*fos*, c-*jun* genes transcription. Thus, AP-1 is involved in APP gene regulation. NF-κB binds DNA as an heterodimer of two polypeptides of 50 and 65 kDa, referred as to p50 and p65 NF-κB. Both of these can also bind DNA with slightly distinct affinity and specificity. Both p65 and p50 have a highly homologous region of 300 amino acids, the NRD domain, which explains the homo- or heterodimerization property and similar DNA binding. Transfection assays with a reporter gene under control of kB motifs have shown that the unique C-terminal 100 amino acids of p65 are sufficient for the transactivating effect while the NRD domains of both peptides are inactive in transactivation. pP50 homodimers which lack this C terminal sequence have been proposed to have a negative regulatory role and to play a physiological role during T cell activation. In most cell types, NF-κB is present in cytoplasm, in an inactive complexed form with a third peptide, I-kB [48, 49]. NF-κB activation leads to dissociation of I-kB from p65; NF-κB can translocate to the nucleus. The processes that activate NF-κB are still unclear. Protein kinase C, reactive oxygen intermediates, and ceramides may be involved in this activation.

C/EBP family members are all leucine-zipper DNA binding protein. C terminal region, with the basic domain and leucine-zipper domain, is highly conserved. Thus the members of the C/EBP family can form either homo- or heterodimers which can bind to the DNA sequences. CHOP10 is an exception, does not bind to DNA, and is believed to be an inhibitor. The amino-terminal region is variable and contains information essential for transcriptional activation. Three members of this family are particularly involved in the hepatic APR: C/EBPα, C/EBPβ (NF-IL-6), and C/EBPδ. C/EBPβ has been shown to be constitutively present in the nucleus. Although neither its mRNA level nor its protein level changes, its transcriptional activating potential increases when IL-6 is added to the medium. C/EBPβ has been shown to be phosphorylated differently with serine or threonine, in different cell types, and through different pathways (calcium/calmodulin-dependent kinase, M-associated protein kinase, protein kinase C). C/EBPβ can interact physically with NF-kB, which can synergize with C/EBP factors in activating promoters containing single or multiple C/EBP binding sites [50, 51]. This may explain the synergistic activation of the expression of some APP genes by IL-1 and IL-6. Thus the two pathways of APP gene activation may be linked.

Recently a second transactivating protein for type 2 AAP genes has been discovered, APRF, or IL-6 responsive element binding protein. APRF binds to the acute-phase responsive element present in the promoter of some APPs such as human and rat α_2-macroglobulin, human β-fibrinogen, haptoglobin, and α_1-acid glycoprotein. IL-6 – in addition to cilliary neurotropic factor, LIF, oncostatin M, and IL-11 – have been shown to trigger the APRF phophorylation in the cytoplasm and its migration to the nucleus.

Resolution of the Acute-Phase Reaction

In most cases APR subsides over 24–48 h, and the organism returns to normal functions over the next several days. The short half-life of most of the cytokines and their mRNA could explain this fact. The secretion of most of the cytokines is inhibited by glucocorticosteroids. The negative regulatory effect of corticosteroids is part of an important negative feedback loop in which IL-1 induces the release of adrenocorticotropic hormone from the hypothalamic pituitary axis. Cytokine-binding inhibitors have been discovered. These exert their action by binding free cytokines in solution, precluding and competing its specific receptor binding. These soluble receptors represent the truncated forms of the membrane forms and arise by proteolitic cleavage of the membrane receptors on the cell surface (shedding). Their affinities for their ligand are usually comparable to those of the membrane receptors. Two distinct TNF-soluble receptors have been identified a myeloid cell type receptor and an epithelial cell type receptor. These TNF binding proteins have been found in normal human urine and in the serum of patient suffering of various form of cancer but not in normal human serum. These proteins are able to inhibit binding and activity of TNF in vitro. An IL-6 soluble receptor has also been identified. In contrast to TNF, however, IL-6/IL-6Rs can interact with gp130 on the cell surface and activate the transduction pathways as a normal ligand-receptor interaction. In addition to soluble receptors, some proteins have been reported to be able to bind certain cytokines and in some cases to inhibit their activity. Uromodulin, or Tamm-Horsfall protein, is a glycoprotein synthesized in the kidney that can bind to IL-1α, IL-1β, and TNFα. It was first discovered in the urine of pregnant women. Two other peptides that can bind TNF have also been identified in urine. As we have noted above, IL-1 and TNF can bind preferentially to some reacted forms of α_2-macroglobulin. Although their activity is partially preserved, they can be rapidly removed from circulation by receptor-mediated endocytosis by various cells such as hepatocytes, fibroblasts, and macrophages. On the other hand, IL-6 binds to native forms of α_2-macroglobulin which may protect it from degradation by trypsin or cathepsin G and from glomerular filtration.

Some other proteins, called receptor antagonists, bind specifically to a cytokine receptor but themselves lack biological activities. They compete with the biologically active cytokine for binding to a common receptor. The only receptor-antagonist identified is the IL-1 receptor antagonist (IL-1ra). IL-1ra is produced by monocytic cells. IL-1ra possesses some structural similarities with IL-1β (26% homology). It binds to IL-1 receptor inhibiting the binding and subsequent signaling of both forms of IL-1. In addition to these molecules, a number of cytokines have been discovered to inhibit the secretion of inflammatory mediators by cells involved in inflammation processes. IL-4 is one of these (reviewed in [52]). IL-4 is produced by T lymphocytes and mast cells and has been shown to inhibit the production of IL-1, TNFα, and prostaglandin E_2 and to upregulate the production of IL-1ra in human monocytes. In hepatocytes, IL-4 inhibits the lipogenesis stimulation by IL-1, IL-6, and TNFα and downregulates the expression of some APP genes [53]. IL-10 (reviewed in [54]) is produced by T cells, B cells, macrophages/monocytes, and keratinocytes and seems to modulate inflammatory response. IL-10 inhibits the production of IL-1, IL-6, IL-8, TNFα, GM-CSF, and G-CSF by LPS-activated monocytes at transcriptional level. IL-10 upregulates IL-1ra production through its mRNA stabilization by LPS-activated polymorphonuclear cells while it inhibits the production of IL-1, TNFα, and IL-8 at the mRNA level. *In vivo* IL-10 has been shown to prevent lethality in experimental endotoxemia-induced septic lethal shock. To our knowledge, the direct effect of IL-10 on APP production has not yet been studied.

Conclusion

The APR must be considered as a physiological reaction following different type of insults in which the liberation of soluble mediators from the site of the aggression influences the expression of protein genes of various cells through highly regulated transduction pathways. The immediate effect of this reaction seems to be host protection. The secretion of opsonins, such as CRP, and complement components protects the host against the invasion of micro-organisms before inititation of the specific immune response. The secretion of antiproteases and superoxyde anion scavengers prevents wide tissular destructions. However, in some circumstances the release of CRP and complement components may increase the generation of eicosanoids, cytokines and leukocytes, and macrophage-derived oxidants to levels that are pathological to the host. This is why efforts are needed to understand the regulation of this phenomenon. New human recombinant anti-inflammatory cytokines may represent a valuable addition to therapeutic techniques.

References

1. Kushner I (1982) The phenomenon of the acute phase response. Ann N Y Acad Sci 389:39–48
2. Rothlein R, Czajkowski M, O'Neil MM et al (1988) Induction of intercellular adhesion molecule-1 on primary and continuous cell lines by pro-inflammatory cytokines: regulation by pharmacologic agents and neutralising antibody. J Immunol 141:1665–1669
3. Carswell EA, Old L J, Kassel R L, Green S, Fiore N, Williamson B (1975) An endotoxin induced serum factor that cause necrosis of tumors. Proc Natl Acad Sci U S A 72:3666–3673
4. Beutler B, Cerami A (1989) The biology of cachectin/TNF-α a primary mediator of the host response. Annu Rev Immunol 7:625–655
5. Sherry B, Cerami A (1988) Cachectin/tumor necrosis factor exerts endocrine, paracrine, and autocrine contol of inflammatory respons. J Cell Biol 107:1269–1277
6. Dinarello CA (1988) Biology of interleukin-1. FASEB J 2:108–115
7. March CJ, Mosley B, Larsen A, Ceretti DP, Braedt G, Price V, Gillis S, Henney CS, Kronheim SR, Grabstein K et al (1985) Cloning, sequence, and expression of two distinct human interleukin-1 complementary DNAs. Nature 315:641–643
8. Chizzonite R, Truitt T, Killian PM et al (1989) Two high affinity interleukin-1 receptors represent separate gene products. Proc Natl Acad Sci U S A 86:8029–8033
9. Ray A, Tatter SB, Santhanam U et al (1989) Regulation of expression of IL-6: molecular and clinical studies. Ann N Y Acad Sci 557:353–362
10. Kishimoto T (1989) The biology of IL-6. Blood 74:1–10
11. Baumann H, Gauldie J (1994) The acute-phase response. Immunol Today 15:75–80
12. Movat HZ, Burrowes CE, Cybulsky MI, Dinarello CA (1987) Acute inflammation and shwarzmann-like reaction induced by interleukin-1 and tumor necrosis factor. Synergistic action of the cytokines in the induction of inflammation and microvascular injury. Am J Pathol 129:463–476
13. Slungaard A, Vercellotti GM, Walker G, Nelson RD, Jacob HS (1990) Tumor necrosis factor α/cachectin stimulates eosinophil oxidant production and toxicity towards human endothelium. J Exp Med 171:2025–2041
14. Rothstein JL, Lint TF, Schreiber H (1988) Tumor necrosis factor/cachectin. Induction of hemorrhagic necrosis in normal tissue requires the fifth component of complement (C5). J Exp Med 168:2007–2021
15. Busbridge NJ, Grossman AB (1991) Stress and the single cytokine: interleukin modulation of the pituitary-adrenal axis. Mol Cell Endocrinol 82:C209–C214
16. Van der Poll T, Levi M, Büller HR, Van Deventer SJH, Boer JP, Hack CE, Ten Cate JW (1991) Fibrinolitic response to tumor necrosis factor in healthy subjects. J Exp Med 174:729–732
17. Argiles JM, Lopez-Soriano FJ, Wiggins D, Williamson DH (1989) Comparative effects of tumor necrosis factor α (cachectin), interleukin-1 β, and tumor growth on amino acid metabolism in the rat in vivo: absorption and tissue uptake of α-amino[1-^{14}C] isobutyrate. Biochem J 261:357–362
18. Roh MS, Moldawer LL, Eckman LG, Dinarello CA, Bistrian BR, Jeevanandam M, Brennan MF (1986) Stimulatory effect of interleukin-1 upon hepatic metabolism. Metabolism 35:419–425
19. Bereta J, Kurdowska A, Koj A, Hirano T, Kishimoto T, Content J, Fiers W et al (1989) Different preparations of natural and recombinant human interleukin-6 (IFN-β_2, BSF-2) similarly stimulate acute-phase protein synthesis and uptake of α-aminoisobutiric acid by cultured rat hepatocytes. Int J Biochem 21:361–366
20. Bertini R, Bianchi M, Erroi A, Villa P, Ghezzi P (1989) Dexamethasone modulation of in vivo effects of endotoxin, tumor necrosis factor, and interleukin-1 on liver cytochrome P-450, plasma fibrinogen, and serum iron. J Leukoc Biol 46:254–262
21. Koj A (1974) Acute-phase reactants. In: Allison AC (ed) Structure and function of plasma proteins, vol 1. Plenum, New York, pp 73–131

22. Baumann H, Gauldie J (1990) Regulation of the hepatic acute phase plasma protein genes by hepatocyte stimulating factors and other mediators of inflammation. Mol Biol Med 7:147–159
23. Lord KA, Abdollahi A, Thomas SM, DeMarco M, Brugge J S, Hoffman-Liebermann B, Liebermann DA (1991) Leukemia inhibitory factor and interleukin-6 trigger the same immediate early response, including tyrosine phosphorylation upon induction of myeloid leukemia differentiation. Mol Cell Biol 11:4371–4378
24. Tillet WS, Francis T Jr (1930) Serologic reactions in pneumonia with a non-protein somatic fraction of pneumococcus. J Exp Med 52:561–571
25. Anderson JK, Mole JE (1982) Large scale isolation and partial primary structure of human plasma amyloid P component. Ann N Y Acad Sci 389:216–234
26. Pepys MB, Dash TC, Fletsher N et al (1978) Analogues in other mammals and in fish of human plasma proteins, C-reactive protein and amyloid P component. Nature 273:168–170
27. Pepys MB, Baltz ML (1983) Acute-phase proteins with special reference to C-reactive protein and related proteins (pentraxins) and serum amyloid A protein. Adv Immunol 34:141–212
28. Robey FA, Jones KD, Tanaka T, Liu TY (1984) Binding of C-reactive protein to chromatine and nucleosome core particles. A possible physiological role of C-reactive protein. J Biol Chem 259:7311–7316
29. Benditt EP, Eriksen N (1971) Chemical classes of amyloid substance. Am J Pathol 65:231–252
30. Levin M, Pras M, Franklin EC (1973) Immunologic studies of the major nonimmunoglobulin protein of amyloid. Identification of partial characterization of a related serum component. J Exp Med 138:373–380
31. Linke RP, Sipe JD, Pollock PS et al (1975) Isolation of a low molecular weight serum component antigenically related to an amyloid fibril protein of unknown origin. Proc Natl Acad Sci U S A 72:1473–1476
32. Hoffman JS, Ericsson LH, Eriksen N et al (1984) Murine tissue amyloid protein AA. NH_2–terminal sequence identity with only one of two serum amyloid proteins (apoSAA) gene products. J Exp Med 159:641–646
33. Crabtree GR, Kant JA (1982) Organization of the rat fibrinogen gene: alternative mRNA splice patterns produces the gamma A and gamma B (gamma) chains of fibrinogen. Cell 31:159–166
34. Kant JA, Fornace AJ, Saxe D et al (1985) Evolution and organization of the fibrinogen locus on chromosome 4: gene duplication accompanied by transposition and inversion. Proc Natl Acad Sci U S A 82:2344–2348
35. Perlmutter DH, Colten HR (1987) Structure and expression of the complement genes. Pharmacol Ther 34:247–270
36. Ross GD, Medof ME (1985) Membrane complement receptors specific for bound fragments of C3. Adv Immunol 37:217–267
37. Gordon H (1976) The α_2–macroglobulin in rat serum. Biochem J 159:643–650
38. Gehring MR, Shiels BR, Northemann W, de Bruijn MHL, Kan CC, Chain AC, Noonan DJ, Fey GH (1987) Sequence of rat liver α_2–macroglobulin and acute phase control of its messenger RNA. J Biol Chem 262:446–454
39. Gonias SL, Pizzo SV (1983) Chemical and structural modification of α_2–macroglobulin: effects on receptor binding and endocytosis studied in an in vivo model. Ann N Y Acad Sci 421:457–471
40. Strickland DK, Ashcom JD, Williams S et al (1990) Sequence identity between the α_2–macroglobulin receptor and the low density lipoprotein receptor-related protein suggests that this molecule is a multifunctional receptor. J Biol Chem 265:17401–17404
41. Kristensen T, Moestrup SK, Gliemann J et al (1990) Evidence that the newly cloned low density lipoprotein receptor-related protein (LRP) is the α2-macroglobulin receptor. FEBS Lett 276:151–155

42. Kristensen T, Moestrup SK, Gliemann J et al (1990) Evidence that the newly cloned low density lipoprotein receptor related protein (LRP) is the α_2–macroglobulin receptor. FEBS Lett 276:151–155
43. Matsuda T, Hirano T, Nagasawa S, Kishimoto T (1989) Identification of α_2–macroglobulin as a carrier protein for IL-6. J Immunol 142:148–152
44. Borth W, Luger TA (1989) Identification of α_2–macroglobulin as a cytokine binding plasma protein. Binding of interleukin-1b to F α_2–macroglobulin. J Biol Chem 264:5818–5825
45. Wollenberg GK, LaMarre J, Rosendal S et al (1991) Binding of tumor necrosis factor alpha to activated forms of human plasma alpha$_2$ macroglobulin. Am J Pathol 138:265–272
46. Rothshild MA, Orath M, Schreiber SS (1988) Serum albumin. Hepatology 8:385–401
47. Fey GH, Fuller GM (1987) Regulation of acute-phase gene expression by inflammatory mediators. Mol Biol Med 4:323–338
48. Ghosh S, Baltimore D (1991) Activation in vitro of NF-kB by phosphorilation of its inhibitor kI-B. Nature 344:678–680
49. Leclair KP, Blanar MA, Sharp PA (1992) The p50 subunit of NF-kB associates with the NF-IL-6 transcription factor. Proc Natl Acad Sci U S A 89:8145–8151
50. Stein B, Cogswell PC, Baldwin AS (1993) Physical and functionnal association between NF-kB and C/EBP family members: a rel domain-bZIP interaction. Mol. Cell. and Biol. 13:3964–3972
51. Paul WE (1991) Interleukin-4: a prototypic immunoregulatory lymphokine. Blood 77:1859–1870
52. Loyer P, Ilyn G, Razzak ZA, Banchereau J, Dezier JF, Campion JP, Guguen-Guillouzo C, Guillouzo A (1993) Interleukin-4 inhibits the production of some acute-phase proteins by human hepatocytes in primary culture. FEBS Lett 336:215–220
53. Spits H, de Waal Malefyt R (1992) Functional characterisation of human IL-10. Int Arch Allergy Immunol 99:8–15
54. Dinarello CA (1992) Role of interleukin-1 in infectious diseases. Immunol Rev 127:119–145

Effects of Catecholamines on Hepatic Blood Flow: Tissue Oxygenation

A. Meier-Hellmann and K. Reinhart

Introduction

Multiple organ failure is the major cause of death in patients with sepsis [1]. Macro- and microcirculatory maldistribution of blood flow to the tissues and hypoxia of certain organs plays an important role in the pathogenesis of sepsis [2, 3]; in addition, an oxygen extraction deficit is at present the subject of research. Therefore a supranormal oxygen delivery (DO_2) is required to guarantee adequate tissue oxygenation [4, 5]. Sufficient volume substitution and treatment with catecholamines is usually required to achieve a supranormal oxygen delivery and an adequate systemic perfusion pressure. Despite this, the mortality rate of sepsis and the incidence of multiple organ failure remains persistently high [6]. It is obvious that tissue hypoxia may occur in different tissues under the conditions of supranormal DO_2 [2, 4]. Information regarding perfusion and oxygenation in various regions of the body is mandatory for the detection of regional tissue hypoxia. Intensive care regimens, particularly treatment with catecholamines, and their influence on perfusion and oxygenation in various organs and tissues are not fully understood and require further investigation.

Renal and gastrointestinal failure are major complications in sepsis [7–9]. Bacterial translocation from the gut is considered by some authors to induce and maintain sepsis [10–12]. Nelson et al. [13] demonstrated that the critical DO_2 of the small intestine in septic sheep is higher than whole-body DO_2. The achievement of critical whole-body DO_2 does not exclude oxygen supply dependency in a number of tissues [13].

Previous studies have not confirmed the reduction of splanchnic perfusion in sepsis. One reason for this may be that different methods were used measure splanchnic perfusion [14]. Dahn et al. [15] demonstrated that splanchnic perfusion in septic patients increases to the same extent as cardiac output (CO), although splanchnic oxygen consumption (VO_{2hep}) increases to a far greater extent than whole-body oxygen consumption (VO_2). Redistribution of blood flow and relative improvement in hepatic and small intestine perfusion have been reported in experimental settings [16, 17]. The infusion of endotoxin doubled splanchnic blood flow in healthy individuals [18]. We found that 60% of the cardiac output in septic patients is distributed through the splanchnic region, compared to 18%–30% in healthy individuals [19, 20].

Table 1. Effects of catecholamines on receptors

Catecholamine	α_1	α_2	β_1	β_2	DA_1	DA_2
Dobutamine	+	0	+++	++	0	0
Epinephrine	+++	+++	++	+++	0	0
Norepinephrine	+++	+++	++	+	0	0
Dopamine						
0–3 $\mu g\ kg^{-1}\ min^{-1}$	0	+	0	0	+++	++
2–10 $\mu g\ kg^{-1}\ min^{-1}$	+	+	++	+	++	++
>10 $\mu g\ kg^{-1}\ min^{-1}$	++	++	++	+	+	+
Dopexamine	0	0	+	+++	++	+

Table 2. Effects of catecholamines on regional blood flow

Catecholamine	Kidney	Brain	Heart	Splanchnic	Muscle	Skin
Dobutamine	+	+	+	+	++	+
Epinephrine	−/+	+	+	−/+	+/0	−
Norepinephrine	−/+	+	+	−/+	−/0	0
Dopamine						
0–3 $\mu g\ kg^{-1}\ min^{-1}$	+++	+	0	+++	0	0
2–10 $\mu g\ kg^{-1}\ min^{-1}$	++/+	+	+	++/+	0	0
>10 $\mu g\ kg^{-1}\ min^{-1}$	−/+	+	+	−/+	−	−
Dopexamine	+++	+	+	+++	+	+

Little is known about the influence of various catecholamines on regional blood flow and oxygenation [21, 22], and the catecholamines most commonly used in the clinical setting, for example, dobutamine, dopamine, norepinephrine, epinephrine, and dopexamine, have generally been investigated in patients without sepsis or not in septic shock (Tables 1, 2).

Dobutamine

Many authors suggest that due to its β_1-receptor mediated effects dobutamine is the catecholamine of choice for increasing myocardial contractility and achieving supranormal CO and DO_2 levels [23–26]. Animal investigations [27] and clinical studies [4, 28] have shown that an increase in DO_2 induced by dobutamine infusion leads to an increased VO_2. Furthermore, it is hypothesized that the β_2-mediated effects counteract peripheral vasoconstriction and hence improve tissue oxygenation [26]. Under the conditions of adequate volume replacement dobutamine increases cardiac output and DO_2 to a greater extent than dopamine [27]. Vincent et al. [28] demonstrated that infusion of dobutamine at 5 $\mu g\ kg^{-1}\ min^{-1}$ in adequately volume-replaced septic patients increases DO_2 by 29% and VO_2 by 18%. Furthermore they speculated that dobutamine decreases tissue perfusion maldistribution. To our knowledge it is not clear whether dobutamine selectively influences splanchnic or renal perfusion. Improvement in splanchnic oxygenation is supported by the results of Silverman et al. [29] who found that infusion of dobutamine at 5 $\mu g\ kg^{-1}\ min^{-1}$ increased a pathologically reduced gastric mucosal pH (pHi); however, blood transfusion did not affect pHi but did increase DO_2 to a similar extent. In septic pigs Schneider et al. [30] found that adequate volume substitution alone restored a decreased cardiac output and led to redistribution of blood flow in favor of the splanchnic region. The additional infusion of dobutamine did not affect the distribution of blood flow to the kidneys and splanchnic region. Leier et al. [31] found that patients with congestive heart failure who received dobutamine at 7.5 $mg\ kg^{-1}\ min^{-1}$ experienced an increase in CO accompanied by a decrease in splanchnic perfusion; in the same study renal perfusion was increased to a lesser extent than CO. Mousdale et al. [32] measured only a minor increase in renal blood flow under dobutamine and dopexamine. However, different doses of dopamine induced remarkable increases in renal blood flow.

In conclusion, there is some evidence that the use of dobutamine to increase DO_2 can lead to an improvement in tissue oxygenation. Whether dobutamine selectively influences regional blood flow is unknown. Management of septic shock with dobutamine alone is often not sufficient to restore adequate blood pressure. Whether the combination of dobutamine and norepinephrine is superior to a treatment with dopamine alone has not yet been investigated.

Norepinephrine

Norepinephrine has only moderate β_1- and β_2-mimetic effects. Experimental models have been shown it to increase splanchnic vascular resistance and to decrease splanchnic blood flow [33]. Norepinephrine has been used in animal studies to induce renal failure [34]. Consequently, norepinephrine is often used only as a last resort when hemodynamic stabilization has not been achieved with other catecholamines [35].

Melchior et al. [36] found that infusion of 0.5 and 1.0 $\mu g\ kg^{-1}\ min^{-1}$ norepinephrine did not change whole-body VO_2 despite increased cardiac index and mean arterial pressure (MAP). They concluded that the beneficial β-mimetic effects of norepinephrine outweigh the possible deleterious effects mediated by α-receptors. In septic dogs Bakker et al. [37] demonstrated an increase of right and left ventricular stroke work index with the infusion of norepinephrine at 0.1–0.2 $\mu g\ kg^{-1}\ min^{-1}$ and an increase in DO_2 and VO_2 with the infusion of norepinephrine at 0.5–1.0 $\mu g\ kg^{-1}\ min^{-1}$. Schneider et al.

[30] recommend norepinephrine over dobutamine for treating right heart failure in septic shock since an increase in diastolic blood pressure improves myocardial perfusion. Schreuder et al. [38] reported that norepinephrine infusion in patients in septic shock increases VO_2 but does not change DO_2. However, it remains unknown whether the increase in VO_2 is caused by an improvement in perfusion pressure, which increased simultaneously from 57 to 75 mmHg. This finding underlines the importance of adequate perfusion pressure and shows that treatment with norepinephrine is indicated if adequate perfusion pressure cannot be maintained with volume substitution and dobutamine or dopamine. In six patients with septic shock in whom hemodynamic stabilization by dopamine or dobutamine infusion was not achieved, DO_2 and VO_2 values were inconsistent when they were treated with norepinephrine [39]. In our own investigations [40] the change in catecholamine treatment from dobutamine to norepinephrine in patients with septic shock led to a decrease in DO_2, but VO_2 remained unchanged. The unchanged VO_2 can be explained by an unchanged tissue oxygenation. However, redistribution of blood flow with norepinephrine treatment alone, undetected by the measurement of the global O_2 transport related variables cannot be excluded.

A number of authors have demonstrated an increase in urine output in patients with septic shock treated with norepinephrine [41–43]. However, it should be emphasized that these patients had markedly decreased blood pressures. In the study of Desjars et al. [41] the infusion of 0.5–1.0 $\mu g\ kg^{-1}\ min^{-1}$ norepinephrine increased MAP from 48 to 62 mmHg. Hesselvik and Brodin [42] reported an increase in MAP from 50 to 69 mmHg after the infusion of 0.5 $\mu g\ kg^{-1}\ min^{-1}$ norepinephrine. The principal mechanism of improved renal function in these studies is the increase in perfusion pressure to an adequate level. Redl-Wenzl et al. [44] reported increased creatinine clearance in 56 patients with septic shock after treatment with norepinephrine. However, in this study the patients also had an inadequate perfusion pressure of 56 mmHg before treatment with norepinephrine. It is unknown whether treatment with norepinephrine has beneficial effects on renal function in the presence of an adequate perfusion pressure. Fukuoka et al. [45] reported in septic patients with lactate levels in a normal range that a norepinephrine infusion increased urine output and did not change creatinine clearance, whereas in patients with elevated lactate levels norepinephrine had no effect on urine output but decreased creatinine clearance.

Findings concerning the effects of norepinephrine on splanchnic perfusion are inconsistent. Bersten et al. [46] compared the effects of various catecholamines on regional blood flow in septic and nonseptic sheep. The nonseptic animals showed a redistribution of blood flow to the heart and away from brain, kidneys, liver, and pancreas with norepinephrine, dobutamine, dopamine, dopexamine, or salbutamol treatment as measured by microsphere technique. A deleterious effect of norepinephrine on splanchnic perfusion was demonstrated in this study. In contrast, no such redistribution of regional blood flow was found in septic animals. The authors explained these findings by a decreased sensibility of α-receptors and a direct sepsis-associated vasodilation [47–51].

Schneider et al. [30] infused pigs with *Escherichia coli* and stabilized the animals initially only by volume substitution. Further treatment with dobutamine or norepinephrine did not affect the distribution of the given volume. Measurements by radiolabeled erythrocytes showed a decreased pooling of blood in the legs after norepinephrine infusion. Breslow et al. [52] infused pigs with *E. coli* and maintained pulmonary capillary wedge pressure within the normal range by infusing isotonic solutions. Under these conditions norepinephrine, dopamine, and phenylepinephrine given in doses to achieve MAP of 75 mmHg were without selective effects on regional blood flow.

Our own investigation [53] in patients with septic shock have shown a rapid decrease in hepatic venous O_2 saturation ($ShvO_2$) as the result of changing the catecholamine management from dobutamine to norepinephrine, whereas the mixed venous O_2 saturation (SvO_2) decreased only slightly. The difference between SvO_2 and $ShvO_2$ therefore increased significantly. This effect may be explained due to a decrease in splanchnic perfusion mediated by intestinal α-receptors [33, 54, 55]. This finding is in contrast to the reported animal studies [46] which demonstrated unchanged splanchnic perfusion with norepinephrine. In our study the selective decrease in $ShvO_2$ could be a marker for a deterioration in splanchnic oxygenation. To demonstrate this interpretation of a changed $ShvO_2$ the splanchnic blood flow has to be measured and the DO_2 and VO_2 in the splanchnic region have to be calculated.

We measured the splanchnic blood flow in ten patients with septic shock by the green dye dilution technique. After changing the catecholamine treatment from a combination of dobutamine and norepinephrine to norepinephrine alone we observed the previously reported changes in SvO_2 and $ShvO_2$. Treatment with norepinephrine alone led to a parallel decrease in splanchnic perfusion and cardiac output. The relationship between splanchnic perfusion and cardiac output remained unchanged. Furthermore, splanchnic VO_2 remained unchanged. Due to the decrease in splanchnic DO_2 hepatic venous O_2 extraction increased. Therefore situations in which there is a further decrease in splanchnic DO_2 entail an increased risk that norepinephrine treatment alone can cause splanchnic VO_2 to become supply dependent.

In conclusion, it should be pointed out that treatment with vasopressors is often indispensable in the therapeutic management of septic shock. Provided that the DO_2 is in a supranormal range, treatment with norepinephrine alone shows no negative effects on tissue oxygenation. Animal studies have shown that the negative effect of norepinephrine on splanchnic perfusion is seen only in nonseptic conditions.

Epinephrine

Epinephrine is advocated in the treatment of septic shock [56–58] because it leads to a β-receptor mediated increase in cardiac output and an α-receptor mediated increase in perfusion pressure. Low doses of epinephrine have predominantly β-mimetic effects, and only at higher doses do the alpha mimetic effects become predominant [59]. Bollaert et al. [56] investigated the effects of epinephrine at 0.5–1.0 $\mu g\ kg^{-1}\ min^{-1}$ in 13 patients with septic shock that could not stabilized by 15 $\mu g\ kg^{-1}\ min^{-1}$ dopamine. Epinephrine increased MAP, systemic vascular resistance, and cardiac output in these patients: furthermore, the DO_2 and VO_2 increased. Elevated arterial lactate levels were interpreteted as a metabolic effect due to epinephrine. Moran et al. [60] reported an increase in DO_2 from 481 to 531 $ml\ min^{-1}\ m^{-2}$ and an increase in VO_2 from 165 to 193 $ml\ min^{-1}\ m^{-2}$ with an epinephrine infusion in doses up to 18 μg/min in patients with septic shock. Depending upon the dose-dependent effects of epinephrine, either a reduction [61–63] or an improvement [64] in splanchnic perfusion has been reported in animal studies. Little is known about the effects of epinephrine on splanchnic and renal perfusion in septic shock states.

In own investigations, changing the catecholamine treatment from a combination of dobutamine and norepinephrine to epinephrine alone decreased $ShvO_2$ but did not change SvO_2, and the difference between the two catecholamine regimens was therefore significantly increased [53]. Dahn et al. [65] reported a parallel decrease in $ShvO_2$ and SvO_2 under hypoxia. In contrast, 33 patients undergoing liver surgery showed decreased $ShvO_2$ but unchanged SvO_2 after skin incision [66]. The authors explained this discrepancy as selective decrease in splanchnic perfusion mediated by endogeneous catecholamines liberated following the skin incision. The selective decrease in $ShvO_2$ seems to be a marker for deterioration in splanchnic oxygenation. We measured splanchnic perfusion, DO_2, and VO_2 in three patients with septic shock treated with a combination of dobutamine and norepinephrine. After a change of treatment to epinephrine alone, titrated to achieve the same MAP as before, DO_2 and VO_2 were unchanged, but epinephrine decreased the splanchnic perfusion in the presence of an unchanged cardiac output. The decreased splanchnic DO_2 led to decreased splanchnic VO_2. These extreme changes in splanchnic oxygenation were not detected by whole-body DO_2 and VO_2, underlying the importance of measuring DO_2 and VO_2 in specific regions. Furthermore, increased lactate levels and decreased pHi indicate tissue hypoxia in the splanchnic region.

Due to the lack of information regarding epinephrine in the treatment of septic shock no conclusion is possible. The preliminary results of our investigation demonstrate a decrease in splanchnic perfusion and oxygenation accompanied by signs of severe tissue hypoxia. If further investigations also demonstrate these negative effects, the use of epinephrine in patients with septic shock cannot be recommended.

Dopamine

Dopamine has effects on α-, β-, and dopaminergic receptors. The dopaminergic receptors can be divided into presynaptic (DA_1) and postsynaptic

(DA_2) receptors. Stimulation of DA_1 receptors causes renal and splanchnic vasodilation, and that of DA_2 receptors blocks norepinephrine output from sympathetic nerves [67].

Dopaminergic receptors are found in renal, mesenteric, and hepatic vessels [67]. The effects of dopamine are dose dependent. In doses up to 3 $\mu g\ kg^{-1}\ min^{-1}$ it selectively stimulates the dopaminergic receptors, up to 5 $\mu g\ kg^{-1}\ min^{-1}$ β-receptors, and over 5 $\mu g\ kg^{-1}\ min^{-1}$ α-receptors.

Vincent and Preiser [68] advocate dopamine as the drug of choice in the primary management of decreased MAP in patients in septic shock. However, other authors maintain that the α-mimetic effects of dopamine are a disadvantage because a previously disturbed microcirculation could be further compromised [26, 27]. In 437 critical care patients Shoemaker et al. induced an increase of DO_2 and VO_2 more frequently with dobutamine than with dopamine [35]. In brain-dead patients treatment with dopamine to elevate MAP to the normal range caused a lower mitochondrial redox state in liver cells than in patients with a lower MAP who were not treated with dopamine [69]. This effect of dopamine can be interpreted as a hypoxia-mediated organ failure.

Dopamine in low doses is often recommended as an additional drug for improving renal function and splanchnic perfusion. Nevertheless it has not been confirmed that treatment with low-dose dopamine is beneficial to patients in septic shock.

In animals and healthy humans dopamine infusion has been demonstrated to increase inulin clearance, renal plasma flow, and sodium excretion [67]. Mousdale et al. [32] demonstrated in healthy humans that dopamine in doses of 2.5, 5.0, and 10.0 $\mu g\ kg^{-1}\ min^{-1}$ induces a significantly higher renal plasma flow than dobutamine in a similar dose and dopexamine in doses of 1.0, 2.0, and 4.0 $\mu g\ kg^{-1}\ min^{-1}$. The effects of dopamine on renal function in patients with septic shock are not completely understood. It is not known whether dopamine influences the incidence of renal failue in septic shock patients.

Dopamine improved a decreased hepatic blood flow in dogs ventilated with positive end-expiratory pressure [70]. In septic rats dopamine induced a selective improvement of hepatic blood flow [71]. An increase in hepatic blood flow does not guarantee improved splanchnic oxygenation because a redistribution of blood flow in the splanchnic area may lead to a deterioration in tissue oxygenation. Giraud et al. [54] infused dogs with dopamine and measured an increase in blood flow in the superior mesenteric artery and in the muscularis of the gut with a decreased blood flow to the gut mucosa accompanied by a decrease in splanchnic VO_2. The improvement in global blood flow measured in a major artery seems to be accompanied by deterioration in blood flow and oxygenation in other tissues such as the gut mucosa. Pawlik et al. [72] demonstrated that intra-arterial infusion of dopamine at 1 $\mu g\ kg^{-1}\ min^{-1}$ in dogs caused splanchnic vasoconstriction with reduction of blood flow and splanchnic VO_2. In the same investigation epinephrine also led to a reduction in splanchnic blood flow but did not affect splanchnic VO_2. Therefore the authors suggested that dopamine is a drug with deleterious side effects on splanchnic oxygenation. In contrast to these findings, Kullmann et al. [73] showed increased blood flow to the intestinal organs and improved perfusion of the mucosa and submucosa in rabbits, as measured by microsphere technique. Leier [31] reported in congestive heart failure patients that 3 $\mu g\, kg^{-1}\, min^{-1}$ dopamine significantly increased renal blood flow and did not change splanchnic blood flow. In nonseptic pigs Roytblat et al. [74] measured a dose-independent increase in splanchnic DO_2 and a dose-dependent increase in splanchnic VO_2 after dopamine infusion. Since higher doses of dopamine increase splanchnic VO_2 to a greater extent than splanchnic DO_2 the authors conclude that dopamine does not improve splanchnic oxygenation and explained the increased splanchnic VO_2 as a calorigenic effect of dopamine [75–77].

In own studies the addition of 2.8–3.0 $\mu g\, kg^{-1}\, min^{-1}$ dopamine to norepinephrine increased $ShvO_2$ more than SvO_2. The difference between the two declined after dopamine infusion. On the one hand, this finding correlates with the reported improvement in splanchnic blood flow under dopamine [70, 78, 79]; on the other, there is no guarantee that the increased $ShvO_2$ is not the result of decreased splanchnic VO_2. In some individuals we observed increased splanchnic VO_2 with low-dose dopamine when dopamine was added to norepinephrine, or dopamine was added to dobutamine. However, further investigations are required to confirm the effects of dopamine.

It has not yet been confirmed whether management with a higher dose of dopamine alone compared to the combination of dobutamine and norepinephrine is of advantage in treating septic shock. The improvement in renal function is an argument for the use of low-dose dopamine, although under the conditions of septic shock there are only few and nonequivocal studies. Animal studies have reported deleterious effects of dopamine on splanchnic oxygenation. One study demonstrated an increased splanchnic VO_2 after dopamine infusion while it remained unclear whether the increased VO_2 was the result of improved oxygenation or only the calorigenic effect of dopamine.

Dopexamine

The relatively new catecholamine dopexamine has predominantly β_2- and dopaminergic receptor activity. Many animal and human studies have described the value of dopexamine in heart failure [80–83]. Tan et al. [84] demonstrated a direct inotropic effect of dopexamine in patients with low-output cardiac failure.

To date there have been only few studies examining the effect of dopexamine in patients with septic shock. Colardyn et al. [85] reported increased cardiac index and heart rate and decreased systemic vascular resistance; unfortunately, data on the DO_2 and VO_2 were not mentioned. Cain and Curtis [86] observed no difference between the mesenteric venous blood flow in a group of septic dogs treated with dopexamine and that in a control group without catecholamine treatment. The gut of the dopexamine-treated animals produced less lactate, and the authors concluded that dopexamine causes gut mucosa to be preferentially perfused. Biro et al. [87] measured a relatively higher skeletal muscle and gastric blood flow in nonseptic dogs with dopexamine in comparison to dobutamine. In dogs in hemorrhagic shock Lokhandwala and Jandhyala [88] demonstrated an increase in splanchnic perfusion to 40%–50% from baseline after blood retransfusion, whereas splanchnic perfusion increased to 80%–90% of baseline when dopexamine was added to the transfusion.

In our studies the addition of 4.0 $\mu g\ kg^{-1}\ min^{-1}$ dopexamine did not affect the difference between SvO_2 and $ShvO_2$. This finding is in agreement with that of Leier [31] who reported in patients with heart failure that increase of cardiac output accompanied that in splanchnic perfusion.

To date, little is known concerning the effects of dopexamine on renal blood flow in patients with septic shock. Dopexamine increases urine output in both healthy and patients with heart failure [89]. In comparison with dopamine, there is no further increase in urine output in patients undergoing hepatic surgery [90]. Stephan et al. [91] demonstrated a more profound decrease in renal vascular resistance under dopamine than under dopexamine in patients with cardiac surgery. The extent of the increase in renal blood flow was the same as that in cardiac output, and the authors concluded that no selective dopaminergic effect of dopexamine was proven. Magrini et al. [92] reported a greater increase in renal blood flow to a higher extent than in cardiac output among patients with hypertension. Mousdale et al. [32] demonstrated that dopamine in doses of 2.5, 5.0, and 10.0 $\mu g\ kg^{-1}\ min^{-1}$ increased renal plasma flow more than dopexamine in doses of 1.0, 2.0, and 4.0 $\mu g\ kg^{-1}\ min^{-1}$.

In conclusion, there is some evidence that dopexamine improves splanchnic oxygenation and renal function, but further studies are required to confirm the value of dopexamine in treating septic shock patients and to compare its effects with those of low-dose dopamine.

Conclusion

Only few, controversial studies exist regarding effects of the different catecholamines on regional blood flow, and no clear recommendation for a specific catecholamine regimen in septic shock can therefore be made. Furthermore, it is unknown whether the choice of specific catecholamine in the treatment of septic shock affects the patient's outcome. It is suggested that the increase in DO_2 to a supranormal range is associated with an improved patient outcome [93–95]. A fundamental step in the treatment of septic shock is adequate volume substitution. In our studies we have demonstrated that consequent volume treatment improves pHi [96]. In most patients the use of vasopressors is indispensable because adequate hemodynamic perfusion pressure is mandatory. The negative effects of vasopressors on splanchnic perfusion are known from studies carried out under nonseptic conditions. Norepinephrine and dopamine in doses of 10 $\mu g\ kg^{-1}\ min^{-1}$ in septic animals are without negative effects on splanchnic perfusion. Preliminary results show decreased splanchnic oxygenation in patients with septic shock treated with epinephrine. Catecholamines with β-mimetic effects are often used to increase DO_2. The question of whether dobutamine or dopamine should be primarily used in treatment of septic shock and cannot be yet answered. Whether treatment with low-dose dopamine or dopexamine actually improves renal function and splanchnic oxygenation is the subject of ongoing studies.

References

1. Fiddian-Green RG (1991) Role of gut in shock and resuscitation. In: Stoutenbeck CP, van Saene HKF (eds) Infection and the anesthetist, vol 5 (1). Bailliere Tindall/Saunders, London, pp 75–99
2. Schumacker P, Cain S (1987) The concept of a critical oxygen delivery. Intensive Care Med 13:223–229
3. Sibbald WJ, Bersten A, Rutledge FS (1989) The role of tissue hypoxia in multiple organ failure. In: Reinhart K, Eyrich K (eds) Clinical aspects of oxygen transport and tissue oxygenation. Springer, Berlin Heidelberg New York, pp 102–114
4. Gilbert EM, Haupt MT, Mandanas RY, Huaringa AJ, Carlson RW (1986) The effect of fluid loading, blood transfusion, and catecholamine infusion on oxygen delivery and consumption in patients with sepsis. Am Rev Respir Dis 134:873–878
5. Kaufman BS, Rackow EC, Falk JL (1984) The relationship between oxygen delivery and consumption during fluid resuscitation of hypovolemic and septic shock. Chest 85(3):337–340
6. Parker MM, Parillo JF (1983) Septic shock. Hemodynamics and pathogenesis. JAMA 250:3324–3327
7. Fry DE, Pearlstein L, Fulton RL (1980) Multiple-system organ failure: the role of uncontrolled infection. Arch Surg 115:136–140
8. Lucas CE (1976) The renal response to acute injury and sepsis. Surg Clin North Am 56:953–975
9. Tran DD, Groeneveld ABJ, Van der Meulen J et al (1990) Age, chronic disease, sepsis, organ system failure, and mortality in a medical intensive care unit. Crit Care Med 18:474–479
10. Deitch EA, Berg R, Specian R (1987) Endotoxin promotes the translocation of bacteria from the gut. Arch Surg 122:185–190
11. Marshall JC, Christou NV, Horn R et al (1988) The microbiology of multiple organ failure: the proximal GI tract as an occult reservoir of pathogens. Arch Surg 123:309–315
12. Meakins JL, Marshall JC (1989) The gut as the motor of multiple system organ failure. In: Marston A, Bulkley GB, Fiddian Green RG et al (eds) Splanchnic ischemia and multiple organ failure. Mosby, St Louis, pp 339–348
13. Nelson DP, Samsel RW, Wood LD, Schumacker PT (1988) Pathologic supply dependence of systemic and intestinal O2 uptake during endotoxemia. J Appl Physiol 64:2410–2419
14. Dahn MS, Lange P, Lobdell K, Hans B, Jacobs LA, Mitchel RA (1989) Hepatic blood flow and splanchnic oxygen consumption measurements in clinical sepsis. Surgery 107:295–301
15. Dahn MS, Lange P, Lobdell K, Hans B, Jacobs LA, Mitchell RA (1987) Splanchnic und total body oxygen consumption differences in septic and injured patients. Surgery 101:69–80

16. Lang CH, Bagby GJ, Ferguson JL, Spitzer JJ (1984) Cardiac output and redistribution of organ blood flow in hypermetabolic sepsis. Am J Physiol 246 R:331–337
17. Wang P, Ba ZF, Chaudry IH (1991) Hepatic extraction of indocyanine green is depressed early in sepsis despite increased hepatic blood flow and cardiac output. Arch Surg 126:219–224
18. Fong Y, Marano MA, Moldawer LL et al (1990) The acute splanchnic and peripheral tissue metabolic response to endotoxin in humans. J Clin Invest 85:1896–1904
19. Johnson GA, McNamara JJ (1981) Organ ischemia after hemorrhagic shock. Surg Forum 32:24–26
20. Leevy CM, George W, Lesko W et al (1961) Observations on hepatic oxygen metabolism in man. JAMA 178:565–567
21. Brown RS, Carey JS, Mohr PA, Monson DO, Shoemaker WC (1966) Comparative evaluation of sympathomimetic amines in clinical shock. Circulation 34:260–271
22. Sibbald WJ, Calvin JE et al (1983) Concepts in the pharmacologic and nonpharmacologic support of cardiovascular function in critically ill surgical patients. Surg Clin North Am 63:456–465
23. Dhainaut JF, Edwards JD, Grootendorst AF et al (1990) Practical aspects of oxygen transport: conclusions and recommendations of the Round Table Conference. Intensive Care Med 16:179–180
24. Reinhart K, Hannemann L, Kuss B (1990) Optimal levels of O2 delivery in the critically ill. Intensive Care Med 16:149–154
25. Shoemaker WC, Appel PL, Kram HB, Bishop M, Abraham E (1993) Hemodynamic and oxygen transport monitoring to titrate therapy in septic shock. New Horizons 1:145–159
26. Shoemaker WC, Appel PL, Kram HB, Duarte D, Harrier D, Ocampo HA (1989) Comparison of hemodynamic and oxygen transport effects of dopamine and dobutamine in critically ill surgical patients. Chest 96:120–126
27. Vincent JL, Van der Linden P, Domb M, Blecic S, Azimi G, Bernard A (1987) Dopamine compared with dobutamine in experimental septic shock: relevance to fluid administration. Anesth Analg 66:565–571
28. Vincent JL, Roman A, Kahn RJ (1990) Dobutamine administration in septic shock: addition to a standard protocol. Crit Care Med 18:689–693
29. Silverman HJ, Tuma P (1992) Gastric tonometry in patients with sepsis. Effects of dobutamine infusions and packed red blood cell transfusions. Chest 102:184–188
30. Schneider AJ, Groeneveld ABJ, Teule GJJ, Nauta J, Heidendal GAK, Thijs LG (1987) Volume expansion, dobutamine and noradrenaline for treatment of right ventrikular dysfunction in porcine septic shock: a combined invasive and radionuclide study. Circ Shock 23:93–106
31. Leier CV (1988) Regional blood flow responses to vasodilators and inotropes in congestive heart failure. Am J Cardiol 62:86E–93E
32. Mousdale S, Clyburn PA, Mackie AM, Groves ND, Rosen M (1988) Comparison of the effects of dopamine, dobutamine, and dopexamine upon renal blood flow: a study in normal healthy volunteers. Br J Clin Pharmacol 25:555–560
33. Granger DN, Richardson PDI, Kvietys PR, Mortillaro NA (1980) Intestinal blood flow. Gastroenterology 78:837–863
34. Mills LC, Moyer JH, Handley CA (1960) Effects of various sympathicomimetic drugs on renal hemodynamics in normotensive and hypotensive dogs. Am J Physiol 198:1279
35. Shoemaker WC, Appel PL, Kram HB (1991) Oxygen transport measurements to evaluate tissue perfusion and titrate therapy: dobutamine and dopamine effects. Crit Care Med 19:672–688
36. Melchior JC, Pinaud M, Blanloeil Y, Bourreli B, Potel G, Souron R (1987) Hemodynamic effects of continious norepinephrine infusion in dog with and without hyperkinetic endotoxic shock. Crit Care Med 15:687–691

37. Bakker J, Vincent JL (1993) Effects of norepinephrine and dobutamine on oxygen transport and consumption in a dog model of endotoxic shock. Crit Care Med 21:425–432
38. Schreuder WO, Schneider AJ, Groeneveld ABJ, Thijs LG (1989) Effect of dopamine vs. norepinephrine on hemodynamics in septic shock. Emphasis on right ventricular performance. Chest 95:1282–1288
39. Meadows D, Edwards JD, Wilkins RG, Nightingale P (1988) Reversal of intractable septic shock with norepinephrine therapy. Crit Care Med 16:663–666
40. Specht M, Meier-Hellmann A, Hannemann L, Spies C, Wirbelauer C, Heil T, Reinhart K (1993) Effects of dobutamine vs norepinephrine therapy on oxygen supply and oxygen consumption in septic patients. Crit Care Med 21:S276
41. Desjars P, Pinaud M, Potel G, Tasseau F, Touze MD (1987) A reappraisal of norepinephrine in human septic shock. Crit Care Med 15:134–137
42. Hesselvik JF, Brodin B (1989) Low dose norepinephrine in patients with septic shock and oliguria: effects on afterload, urine flow, and oxygen transport. Crit Care Med 17:179–180
43. Martin C, Eon B, Saux P, Aknin P, Gouin F (1990) Renal effects of norepinephrine used to treat septic shock patients. Crit Care Med 18:282–285
44. Redl-Wenzl EM, Armbruster C, Edelmann G, Fischl E, Kolacny M, Wechsler-Fördös A, Sporn P (1993) The effects of norepinephrine on hemodynamics and renal function in severe septic shock states. Intensive Care Med 19:151–154
45. Fukuoka T, Nishimura M, Imanaka H, Taenaka N, Yoshiya I, Takezawa J (1989) Effects of norepinephrine on renal function in septic patients with normal and elevated lactate levels. Crit Care Med 17:1104–1107
46. Bersten AD, Hersch M, Cheung H, Rutledge FS, Sibbald WJ (1992) The effect of various sympathomimetics on the regional circulations in hyperdynamic sepsis. Surgery 112:549–561
47. Baker CH, Wilmoth FR (1984) Microvascular responses to E. coli endotoxin with altered adrenergic activity. Circ Shock 12:165–176
48. Gray GA, Furman BL, Parratt JR (1990) Endotoxin-induced impairment of vascular reactivity in the pithed rat: role of arachidonic acid metabolites. Circ Shock 31:395–406
49. Hollenberg SM, Cunnion RE, Parrillo JE (1992) Effect of septic serum on vascular smooth muscle: in vitro studies using rat aortic rings. Crit Care Med 20:993–998
50. Kato T, Hayashi K, Takamizawa K (1988) Response of femoral arteries to norepinephrine following endotoxicosis. Circ Shock 26:383–390
51. Seaman KL, Greenway CV (1984) Loss of hepatic venous responsiveness after endotoxin in anesthetized cats. Am J Physiol 246:H658–H663
52. Breslow MJ, Miller CF, Parker SD, Walman AT, Traystman RJ (1987) Effect of vasopressors on organ blood flow during endotoxin shock in pigs. Am J Physiol 252:H291–H300
53. Meier-Hellmann A, Hannemann L, Specht M, Schaffartzik W, Spies C, Reinhart K (1992) The relationship between mixed venous and hepatic venous O2 saturation in patients with septic shock . Oxygen transport to tissue. Adv Exp Med Biol 345:701–707
54. Giraud GD, MacCannell KL (1984) Decreased nutrient blood flow during dopamine- and epinephrine-induced intestinal vasodilatation. J Pharmacol Exp Ther 230:214–220
55. Reilly FD, McCuskey RS, von Cilento E (1981) Hepatic microvascular regulatory mechanisms. I. Adrenergic mechanisms. Microvasc Res 21:103–116
56. Bollaert PE, Bauer Ph, Audibert G, Lambert H, Larcan A (1990) Effects of epinephrine on hemodynamics and oxygen metabolism in dopamine-resistant septic shock. Chest 98:949–953
57. Lipman J, Roux A, Kraus P (1991) Vasoconstrictor effects of adrenaline in human septic shock. Anaesth Intensive Care 19:61–65

58. Mackenzie SJ, Kapadia F, Nimmo GR, Armstrong IR, Grant IS (1991) Adrenaline in treatment of septic shock: effects on hemodynamics and oxygen transport. Intensive Care Med 17:36–39
59. Innes IR, Nickerson M (1975) Norepinephrine, epinephrine and the sympathomimetic amines. In: Goodman LD, Gilman A (eds) The pharmalogical basis of therapeutics. MacMillan, New York, pp 483–504
60. Moran JL, O'Fathartaigh MS, Peisach AR, Chapman MJ, Leppard P (1993) Epinephrine as an inotropic agent in septic shock: a dose-profile analysis. Crit Care Med 21:70–77
61. Haddy FJ, Chou CC, Scott JB et al (1967) Intestinal vascular responses to naturally occurring vasoactive substances. Gastroenterology 52:444–451
62. Haddy FJ, Molnar JI, Borden CW et al (1962) Comparison of direct effects of angiotensin, and other vasoactive agents on small and large blood vessels in several vascular beds. Circulation 25:239–246
63. Pawlik W, Shepherd AP, Jacobson ED (1975) Effects of vasoactive agents on intestinal oxygen consumption and blood flow in dogs. J Clin Invest 56:484–490
64. Greenway CV, Lawson A (1966) The effects of adrenaline and noradrenaline on venous return and regional blood flow in the anesthetized cat with special reference to intestinal blood flow. J Physiol (Lond) 187:579–595
65. Dahn MS, Lange P, Jacobs LA (1988) Central mixed and splanchnic venous oxygen saturation monitoring. Intensive Care Med 14:373–378
66. Kainuma M, Fujiwara Y, Kimura N, Shitaokoshi A, Nakashima K, Shimado Y (1991) Monitoring hepatic venous hemoglobin oxygen saturation in patients undergoing liver surgery. Anesthesiology 74:49–52
67. Goldberg LI (1972) Cardiovascular and renal actions of dopamine. Potential clinical applications. Pharmacol Rev 24:1–29
68. Vincent JL, Preiser JC (1993) Inotropic agents. New Horizons 1:137–144
69. Nakatani T, Isihkawa Y et al (1991) Hepatic mitochondrial redox state in hypotensive brain-dead patients and an effect of dopamine administration. Int Care Med 17:103–107
70. Johnson DJ, Johannigman JA, Branson RD, Davis K, Hurst JM (1991) The effect of low dose dopamine on gut hemodynamics during PEEP ventilation for acute lung injury. J Surg Res 50:344–349
71. Townsend MC, Schirmer WJ, Schirmer JM, Fry DE (1987) Low-dose dopamine improves effective hepatic blood flow in murine peritonitis. Circ Shock 21:149–153
72. Pawlik W, Shepherd AP, Mailman D, Shanbour LL, Jacobson ED (1976) Effects of dopamine and epinephrine on intestinal blood flow and oxygen uptake. In: Grote J, Reneau D, Thews G (eds) Oxygen transport to the tissue. II. Plenum, New York, pp 511–516
73. Kullmann R, Breull WR, Reinsberg J, Wassermann K, Konopatzki A (1983) Dopamine produces vasodilation in specific regions and layers of the rabbit gastrointestinal tract. Life Sci 32:2115–2122
74. Roytblat L, Gelman S, Bradley EL, Henderson T, Parks D (1990) Dopamine and hepatic oxygen supply-demand relationship. Can J Physiol Pharmacol 68:1165–1169
75. Jackson LK, Key BM, Cain SM (1982) Total and hindlimb O2 uptake and blood flow in hypoxic dogs given dopamine. Crit Care Med 10:327–331
76. Regan CJ, Duckworth R, Fairhurst JA, Maycock PF, Frayn KN, Campbell IT (1990) Metabolic effects of low-dose dopamine infusion in normal volunters. Clin Sci 79:605–611
77. Ruttimann Y, Chidero R et al (1989) Effects of dopamine on total oxygen consumption and oxygen delivery in healthy men. Am J Physiol 257:E541-546
78. Lundberg J, Lundberg D, Norgren L, Ribbe E, Thörne J, Werner O (1990) Intestinal hemodynamics during laparotomy: effects of thoracic epidural anesthesia and dopamine in humans. Anesth Analg 71:9–15
79. Winsö O, Biber B, Martner J (1985) Does dopamine suppress stress-induced intestinal and renal vasoconstriction? Acta Anaesthesiol Scand 29:508–514

80. Baumann G, Gutting M, Pfafferott C, Ningel K, Klein G (1988) Comparison of acute haemodynamic effects of dopexamine hydrochloride, dobutamine and sodium nitroprusside in chronic heart failure. Eur Heart J 9:503–512
81. De Marco T, Kwasman M, Au D, Chatterjee K (1988) Dopexamine hydrochloride in chronic congestive heart failure with improved cardiac performance without increased metabolic cost. Am J Cardiol 62:57C–62C
82. Parratt JR, Wainwright CL, Fagbemi O (1988) Effect of dopexamine hydrochloride in the early stages of experimental myocardial infarction and comparison with dopamine and dobutamine. Am J Cardiol 62:18C–23C
83. Svenson G, Strandberg LE, Lindvall B, Erhardt L (1988) Haemodynamic response to dopexamine hydrochloride in postinfarction heart failure: lack of tolerance after continuous infusion. Br Heart J 60:489–496
84. Tan LB, Littler A, Murray RG (1987) Beneficial haemodynamic effects of intravenous dopexamine in patients with low-output heart failue. J Cardiovasc Pharmacol 10:280–286
85. Colardyn FC, Vandenbogaerde JF, Vogelaers DP, Verbeke JH (1989) Use of dopexamine hydrochloride in patients with septic shock. Crit Care Med 17:999–1003
86. Cain SM, Curtis SE (1991) Systemic and regional oxygen uptake and delivery and lactate flux in endotoxic dogs infused with dopexamine. Crit Care Med 19:1552–1560
87. Biro GP, Douglas JR, Keon WJ, Taichman GC (1988) Changes in regional blood flow distribution induced by infusions of dopexamine hydrochloride or dobutamine in anesthetized dogs. Am J Cardiol 62:30C-36C
88. Lokhandwala MF, Jandhyala BS (1992) Effects of dopaminergic agonists on organ blood flow and function. Clin Intensive Care 3(1) [Suppl]:12–16
89. Baumann G, Felix SB, Filcek SAL (1990) Usefulness of dopexamine hydrochloride versus dobutamine in chronic congestive heart failure and effects on hemodynamics and urine output. Am J Cardiol 65:748–754
90. Gray PA, Bodenham AR, Park GR (1991) A comparison of dopexamine and dopamine to prevent renal impairment in patients undergoing orthotopic liver transplantation. Anaesthesia 46:638–641
91. Stephan H, Sonntag H, Henning H, Yoshimine K (1990) Cardiovascular and renal haemodynamic effects of dopexamine: comparison with dopamine. Br J Anaesth 65:380–387
92. Magrini F, Foulds RA, Roberts N, Macchi G, Mondadori C, Zanchetti A (1988) Renal hemodynamic effects of dopexamine hydrochloride. Am J Cardiol 62:53C–56C
93. Shoemaker WC, Appel PL, Kram HB (1988) Tissue oxygen debt as a determinant of lethal and nonlethal postoperative organ failure. Crit Care Med 16:117–120
94. Shoemaker WC, Appel PL, Kram HB, Waxman K, Lee TS (1988) Prospective trial of supranormal values of survivors as therapeutic goals in high risk surgical patients. Chest 94:1176–1186
95. Tuchschmidt J, Fried J, Swinney R et al (1989) Early hemodynamic correlations of survival in patients with septic shock. Crit Care Med 17:719–723
96. Hannemann L, Meier-Hellmann A, Specht M, Spies C, Reinhart K (1993) O2-Angebot, O2-Verbrauch und Mucosa pH-Wert des Magens. Anaesthesist 42:11–14

Liver-Lung Interactions in Critical Illness

G.M. Matuschak

Introduction

Though acknowledged to have a central role in host defense homeostasis as well as immunological, biochemical, and metabolic regulation, the liver has not been recognized as pivotal to outcome in the acute respiratory distress syndrome (ARDS) for several reasons. Patients with liver disease have often been excluded from study, and hepatic dysfunction has often been nonspecifically defined by "liver function tests" [1]. The liver is not as accessible for study as the lung and other organs, and acute liver dysfunction is not as immediately evident as is acute lung injury [2, 3]. We have proposed an expanded conception of sepsis- and trauma-related ARDS as the central pulmonary manifestation of a generalized disorder of immunoregulation; the pathogenesis and resolution of lung injury are linked to more fundamental derangements in systemic host defense [2–8]. This reorientation implies that understanding the pathways by which changes in hepatic performance affect pulmonary function in ARDS may have therapeutic utility. Here we examine the thesis that hepatic performance modulated predisposition and resolution of lung injury in ARDS by affecting four interrelated elements of host defense: (a) control of systemic endotoxemia, bacteremia, and vasoactive byproducts of sepsis and trauma, (b) regulation of the production and export of endogenous inflammatory mediators by mononuclear phagocytes (Kupffer cells), (c) metabolic inactivation and detoxification of these mediators, and (d) synthesis of acute-phase proteins essential in intermediary metabolism and control of the inflammatory response. As a corollary, we assess clinical and experimental evidence suggesting that alterations in hepatic performance augment lung inflammation and mortality owing to a cytokine:eicosanoid axis of inflammation within the intravascular compartment and lower respiratory tract.

Control of Systemic Endotoxemia, Bacteremia, and Vasoactive By-Products of Sepsis and Trauma

Determinants of Kupffer Cell Uptake and the Gut-Liver-Lung Axis of Cardiopulmonary Homeostasis

Anatomic, physiological, and cellular factors unique to the liver support its regulation of circulating concentrations of microbial products, tissue debris, bioactive particulates, and products of intravascular coagulation [9–13]. Several lines of evidence strongly support the thesis that derangement of one or more elements in this hierarchy initiates and amplifies extrahepatic organ dysfunction associated with sepsis or trauma [7, 8, 14–20]. Even so, the pathogenetic significance of these intrahepatic changes relative to multiple organ dysfunction syndromes and ARDS must be interpreted against the backdrop of hierarchic interactions within a gut-liver-lung axis [7, 21–28].

Hepatic mononuclear phagocytic uptake and detoxification of endotoxin is the major mechanism limiting the magnitude and duration of systemic endotoxemia [9, 14–16, 29–31]. Several key determinants regulate such hepatic clearance mediated by Kupffer cells lining the extensive sinusoidal network, which constitute nearly 80%–90% of the body's reticuloendothelial cell mass. These include:

- The magnitude and kinetics of endotoxin/microbial entry into systemic and portal circuits
  - Systemic vs. peritoneal sepsis
  - Gut-liver interactions: gut mucosal barrier function (altered digestive tract microflora, ischemia-reperfusion injury)
- Liver blood flow
  - Fractional distribution of cardiac output to splanchnic circuit
  - Flow partitioning between portal venous flow and hepatic arterial flow
  - Back pressure to hepatic outflow
  - Intrahepatic distribution of sinusoidal perfusion
  - Transhepatic sinusoidal transit time
- Preexisting liver disease
- Plasma opsonins
  - Nonspecific (fibronectin)
  - Specific (complement peptides, immunoglobulins)
- Receptor-mediated clearance by Kupffer cells
  - CD14 receptor uptake mediated by lipopolysaccharide binding protein (bactericidal permeability-increasing protein-lipopolysaccharide binding protein stoichiometry
  - Fc and complement receptor function
- Non-receptor-mediated endotoxin uptake
- Hepatocytic function and biliary elimination

Besides uptake of endotoxin, microbial pathogens, and sepsis- and trauma-related particulates, Kupffer cell clearance of altered platelets and products of intravascular coagulation protects the lungs and other extrahepatic organs [10, 32–35]. Kupffer cell mediated clearance is subserved by vascular and microcirculatory factors. The splanchnic circulation contains 20%–25% of the systemic blood volume and receives 25% of the cardiac output [36], all of which is gated through the anastomosing hepatic sinusoidal capillaries before returning to the systemic venous circuit. The prolonged transit time and physical contact of blood elements with Kupffer cells optimizes clearance. Delivery of phlogistic substances to the Kupffer cell population varies with two variables: (a) total liver blood flow (Q_L), and (b) Q_L partitioning between portal venous flow (Qpv) comprising up to 75% of Q_L and hepatic arterial flow (Qha) [12]. Although sinusoidal cells receive a mixed arterioportal nutritive supply, differences exist in the pressure-flow and autoregulatory behavior of these circuits [37, 38]. These may have pathogenetic implications under changing conditions of intestinal O_2 delivery [39, 40]. In contrast to the hepatic artery, pressure-flow autoregulation is minimal in the portal circuit [12, 38], such that decreases in Qpv are not offset by corresponding elevations in portal pressure. As Qpv constitutes a large fraction of total Q_L, the hepatic O_2 supply is critically dependent on upstream hemodynamic events [41]. An important feature of hepatic vascular physiology from the standpoint of liver phagocytic clearance and O_2 delivery-uptake is flow reciprocity [37, 42, 43]. By this process reductions in the flow of one circuit are partially compensated by increases in perfusion through the other by changes sensed in osmolarity, pH, gas tensions, or endogenous vasodilators such as adenosine [43]. Although this process increases O_2 delivery during falls in Qpv, compensatory increments by Qha may be inadequate to restore phagocytic clearance of gut-derived substances. The in-series arrangement of mesenteric and portal circuits can thus be viewed as the substrate for occult hemodynamic interactions between the gut and liver [41, 44]. These derive primarily from the sensitivity of the splanchnic circulation to increases in α-adrenoceptor tone during circulatory stress including hypovolemic shock, during which mesenteric inflow is limited by arteriolar vasoconstriction [38, 41]. Sustained sympathoadrenal activation due to inadequate resuscitation or α-agonists in high doses result in protracted splanchnic flow diversion that may compromise the gut mucosal barrier [21, 45, 46]. Whether these otherwise homeostatic effects can be mitigated by dopaminergic receptor stimulation is unclear [47]. In a porcine model of hemorrhagic shock low-dose dopamine ($5\ \mu g\ kg^{-1}\ min^{-1}$) given prior to volume resuscitation improved hepatic as well as subcutaneous and transcutaneous O_2 tensions for the same cardiac output [48].

Nonbacteremic oxidant stress of the gut during mesenteric ischemia-reperfusion activates a complex inflammatory network consisting of activated neutrophils (PMNs), cytokines, and their second messengers. Elements of this network include tumor necrosis factor-α (TNF-α), interleukin (IL) 1, 6, and 8, complement peptides, platelet-activating factor (PAF), and eicosanoid products of the cyclo-oxygenase and lipoxygenase pathways [49, 50]. These may damage the liver owing to its strategic location downstream from the splanchnic reservoir of gram-negative bacteria and their products. Several reports links changes in gut mucosal barrier function owing to reduced O_2 delivery with increases in portal venous endotoxemia and bacteremia [21, 46, 51–53]. Nelson et al. demonstrated that the critical level of gut O_2 delivery needed to maintain enterocyte integrity during endotoxemia was higher than in other organs of the body [54]. Despite reports that shock and organ damage during endotoxemia are mediated chiefly by host-derived proinflammatory mediators such as cytokines, eicosanoids, complement peptides, and reactive intermediates of oxygen and nitrogen [55–64] (see below), lipopolysaccharide (LPS) itself is synergistic with LPS-induced mediators [65].

Hepatic clearance mechanisms for endotoxin or pathogens may be overwhelmed by preexisting liver disease [66, 67], systemic and splanchnic hemodynamic derangements [25, 26, 45, 46], opsonic factor depletion [13], or other changes in the gut microflora [21, 25] (see above). Singly or in combination these may result in "spillover" of microbial products past the liver to secondarily impair lung function [7, 10, 68]. Evidence in support includes the following considerations: (a) patients with end-stage liver disease awaiting liver transplantation are predisposed to ARDS which once initiated, is irreversible despite ventilatory support [8, 69]; (b) established ARDS during acute liver allograft rejection resolves within hours of hepatic retransplantation [70]; (c) pulmonary deposition of *Escherichia coli* is augmented during experimental biliary cirrhosis compared with predominant hepatic uptake in noncirrhotic controls [16]; (d) Decamp et al. reported that hepatic uptake of intraportal colloidal particles and *Pseudomonas* spp. in an ovine model was efficient compared with *E. coli* LPS, a significant fraction of which escaped hepatic clearance to result in lung inflammation [71]; and (e) Matuschak et al. found evidence of enhanced lung uptake of *E. coli* LPS and neutrophilic alveolitis following endotoxemia during steady-state reductions in Q_L in porta-caval shunted rats, compared with those animals having normal Q_L [72].

Data are sparse concerning portal endotoxemia and hepatic vascular responses in terms of real-time changes in endotoxin clearance and functional measures of organ performance. Halvorsen et al. [73] recently showed that portal endotoxemia in a porcine model (LPS infusion of 1 g/h) increased hepatic arterial resistance by nearly 350% and portal venous resistance by 160% in conjunction with incomplete hepatic LPS uptake. Moreover, hepatic endotoxin spillover increased pulmonary vascular resistance. Compared with other regional circulations, most vasoconstriction to portal endotoxemia in this study was limited to the liver and the lungs.

The concept of a multiorgan axis of inflammation coupled by gram-negative bacteremia and/or portal endotoxemia has antecedents. Fine et al. postulated that endogenous endotoxemia of enteric origin accounts for the irreversibility of hemorrhagic shock [74] while Nolan et al. showed that ^{51}Cr-labeled endotoxin is transported across the gut wall [75]. A large database supports this mechanism as a cause of lung injury [76]. Our perspective of these interactions has become more comprehensive in regard to systemic inflammation because of two factors: (a) clarification of the mediators of injury, and (b) recognition that Kupffer cell phagocytic function is one among several important hepatic defense mechanisms.

Q_L During Sepsis and Trauma: Effects of Positive-Pressure Ventilation

Assessment of Q_L during critical illness has been an elusive goal. Elements of flow-dependent hepatic performance can vary independently, making analysis problematic [77–79]. In contrast to measurements of Q_L in healthy subjects by clearance of indocyanine green (ICG) or galactose [80, 81], and by Doppler sonography [82], only limited clinical studies have determined the impact of sepsis and trauma on Q_L [83]. Assessing the dual hepatic vascular input is difficult and usually entails hepatic vein catheterization. The dependence of clearance on hepatocytic function can be also altered independently of sinusoidal perfusion by critical illness or preexisting liver disease ("sick hepatocyte hypothesis") [84]. Dahn et al. found that hepatic ICG extraction decreased by 64% in patients with sepsis, but that Q_L calculated using bolus clearance methods was significantly increased compared with controls [85]. By contrast, parallel increases in the portal fraction of Q_L determined by ICG clearance and Doppler laser flowmetry have been reported early in experimental peritonitis in rats [86]. Q_L and functional responses to nonbacteremic shock or trauma after resuscitation to systemic hemodynamic endpoints such as cardiac output or arterial pressure may nonetheless be accompanied by sustained Q_L reductions and functional impairment [87, 88]. This may reflect no-reflow phenomena secondary to sinusoidal cell swelling, neutrophilic influx, and oxidant-mediated endothelial damage typifying hepatic ischemia-reperfusion injury [89, 90].

Positive-pressure ventilation is used to support impaired lung function but can adversely modulate extrapulmonary organ function [91–97]. The effects of positive-pressure ventilation on Q_L and flow-dependent performance have not been fully defined because of the complexity of forces generated during its application. Phasic inspiratory increases in intrathoracic pressure (ITP) cause parallel increases in the effective hepatic back pressure, whether defined as right atrial pressure (Pra) or hepatic venous pressure (Phv). These could affect hepatic clearance processes by several mechanisms: (a) altering the partitioning of Q_L between Qpv and Qha; (b) increasing the heterogeneity of sinusoidal perfusion owing to generation of intrahepatic closing pressures from hepatic compression by the descending diaphragm; and (c) reducing sinusoidal cell:blood contact time owing to retrograde pressure wave propagation of the inspiratory Pra pulse [93]. Reductions in Qpv by electromagnetic flowmetry during phasic increases in ITP over a respiratory frequency (f) spectrum from 0.4 to 2.5 Hz modulate the outflow pressure in a postsinusoidal flow-limiting segment in which compression of the liver by the diaphragm increases resistance to Qpv in f-dependent manner [93]. However, the pharmacokinetics of ICG showed no f-dependent effects, suggesting that substances with similar hepatic extraction vary with inflow over this f spectrum. No studies have evaluated these issues with respect to Kupffer cell phagocytic function. During positive end-expiratory pressure (PEEP) a variety of changes in Q_L and hepatic performance have been reported. These include reductions in Qpv [94, 95] and decreases in sulfobromophthalein sodium excretion [91]. After accounting for PEEP-related changes in cardiac output we found that phasic respiratory swings in hepatic outflow over single positive-pressure respiratory cycles were not influenced by adding 10 cm H_2O PEEP to intermittent positive pressure ventilation [98]. Moreover, reductions in Q_L during PEEP in two acute canine models were prevented by intravascular volume loading to pre-PEEP cardiac output, and hepatic performance was not impaired as assessed by ICG extraction and clearance. By increasing Q_L gut feeding may also augment Q_L and O_2 delivery despite the otherwise flow-limited state associated with PEEP independently of cardiac output [99]. Important questions remain unanswered concerning the effects of PEEP on the liver. These include whether hemodynamic forces affect mononuclear phagocytic function, and the role of dopaminergic agents including dopexamine on flow-dependent clearance.

Hepatic Production and Export of Inflammatory Mediators

The role of the liver in host defense affecting predisposition to ARDS extends beyond the filtering function of mononuclear phagocytic elements. Hepatic regulation of the production and export of early-acting cytokine mediators after bacteremic and nonbacteremic stimuli is complex, involving interdependent microbial, hemodynamic, and host factors [7, 100, 101]. We focus here on hepatic cytokine production as a second major element of host defense relevant to ARDS, recognizing that other noncytokine mediators such as the recently described "injurin" [102] synergize to cause lung injury. In addition to pathogen-specific characteristics and kinetics of Kupffer cell uptake, several hierarchic mechanisms govern cytokine production: (a) receptor-mediated induction of transcriptional elements regulating cytokine gene expression such as nuclear factor (NF) κB by endotoxin, lipoteichoic acid, or other microbial products [103, 104]; (b) control of cytokine mRNA stability by the untranslated 3' consensus octomer [105]; (c) translational efficiency of mRNA; (d) membrane shedding of soluble receptors that by ligand-specific binding preempt cytokine binding to cell targets [106]; and (e) natural antagonists such as the IL-1 receptor antagonist synthesized in the liver among other sites [107]. LPS-induced cytokine expression by the liver is modulated by a network involving coinduced anti-inflammatory cytokines such as IL-4, IL-6, IL-10, and transforming growth factor-β (TGF-β), eicosanoid species including E-series prostaglandins, PAF, and reactive oxygen species (ROS) [1, 2, 19, 55–64, 108–110].

Hepatic Cytokine Secretory Capacity

What is the capacity of the liver to produce and export inflammatory cytokines into the systemic circulation following gram-negative microbial challenge? The data of Fong et al. [20] show that the liver is a major source of circulating TNF-α early after *E. coli* endotoxemia in humans, as the peak hepatic cytokine efflux was approximately 7 μg. Production of bioactive and antigenic TNF-α during perfusion of rat liver is likewise substantial after intraportal viable *E. coli* [100]. As with TNF-α, IL-6 gene expression increased after gram-negative bacteremia compared with saline controls in that study. Although peak circulating bioactive TNF-α 180 min after an equivalent inoculum of exotoxin C-producing *Staphylcoccus aureus* was similar, TNF-α and IL-6 mRNA levels were significantly less [100]. Despite this there were no pathogen-specific differences in microbial clearance or O_2 uptake. The large secretory output of TNF-α, IL-1, and IL-6 by the liver following LPS challenge has been confirmed [111].

Cytokine Biology in Liver Disease

Hepatic functional impairment modulates cytokine biology in several ways relevant to acute lung injury. Circulating cytokine levels and spontaneous or LPS-induced secretion of TNF-α, IL-1, and IL-6 from mononuclear phagocytes in vitro are increased in patients with liver disease [112–114]. The increases are disease stage-dependent but unrelated to the type or chronicity of liver disease, having been observed in acute viral or alcoholic hepatitis [112, 114], fulminant hepatic failure [115], and cirrhosis [113, 116]. The highest increases in circulating TNF-α, IL-1β, IL-6, γ-interferon, and C-reactive protein (CRP) occur in patient with cirrhosis compared with noncirrhotic hepatic involvement in chronic liver diseases [116]. Liver diseases associated with viral infection may cause ongoing stimulation of hepatic macrophages [117]. Endotoxemia associated with liver disease [66, 67] may also stimulate cytokine production in the absence of viral-induced organ injury. Further, there may be impaired Kupffer cell or hepatocytic clearance of circulating mediators secreted in otherwise normal amounts. These processes may account for the impact of preexisting cirrhosis on circulating cytokine levels and mortality in patients with septic shock. Thus, patients with liver disease had higher serum levels of IL-6 and were at enhanced risk for lethal multiple organ failure compared with noncirrhotic septic patients [118].

Hepatic Cytokine Production in Fungal Sepsis

Infection with *Candida albicans* has increased 20-fold over the past decade [119], with hepatic candidiasis common in ICU patients after trauma, burns, or immunosuppression [120, 121]. Regulation of hepatic cytokine production during candidemia differs compared with gram-negative or gram-positive bacteremia [100, 122]. In contrast to bacteria, intraportal candidemia with human pathogenic strains of viable yeast-phase *C. albicans* elicited the weakest stimulatory responses in perfused rat liver with respect to TNF-α or IL-6 gene expression and TNF-α protein production [100]. These differences between bacterial and fungal infections are likely due to the molecular mimicry of yeast-phase *Candida* spp. adhesion molecules with the mammalian integrin superfamily and other molecules [123], as similar findings have been observed in Kupffer cell cultures [124] in vivo, the magnitude of hepatic and systemic TNF-α production and TNF-α-dependent immunophysiological responses are differentially regulated in conscious rats after lethal *Candida* spp. *vs. E. coli* infection [122]. In neither immunocompetent nor immunosuppressed neutropenic models [125] was TNF-α found to be a pivotal mediator of the acute *Candida* septic shock syndrome with disseminated candidiasis.

Cytokine Kinetics During Nonbacteremic vs. Bacteremic Oxidative Stress of the Liver

Reductions in the hepatic O_2 supply are common during shock caused by hemorrhage, trauma, or gram-negative bacteremia [2, 39, 40, 54, 85, 126–128]. Decreases in intracellular ATP, conversion of hepatic xanthine dehydrogenase to xanthine oxidase [129], and mitochondrial activation [130] generate ROS which modulate the release, binding, and cytotoxicity of cytokines including TNF-α [131]. One mechanism for ROS-related cytokine induction during oxidative hepatic stress is their role as messengers in activating NF-κB which rapidly upregulates cytokine gene expression [103]. Ultimately, changes in cardiovascular function which cause even transient nonbacteremic ischemia-reperfusion and/or hypoxia-reoxygenation of the liver are linked to immunological alterations, manifested by activation of cytokine-dependent inflammatory responses. Colletti et al. demonstrated that 90 min of no-flow hepatic lobar ischemia-reperfusion resulted in focal liver injury in rats [132]. Reperfusion was associated with localized hepatic injury, increased circulating TNF-α, augmented lung microvascular permeability, and enhanced pulmonary leukosequestration, all of which were prevented by pretreatment with anti-TNF-α antiserum. This may explain the enhanced PMN influx into the lungs after hepatic ischemia-reperfusion shown in a similar model by Jaeschke et al. [133], a process linked to cytokine-dependent upregulation of the Mac-1 (CD11b/CD18) PMN adhesion protein [134]. Such upregulation may also explain the increased PMN aggregation in the liver and lungs after soft tissue injury [135]. Ayala and coworkers reported that murine hemorrhagic shock had differential effects on Kupffer cells, with increased TNF, IL-1 and IL-6 release contrasting with reductions in antigen presentation mediated by cyclo-oxygenase metabolites [136]. We recently showed that even brief (e.g., 30 min) periods of global, nonbacteremic hypoxic stress of the intact perfused liver releases TNF-α without increasing

TNF-α mRNA accumulation [137]. This presumably resulted from cleavage at the Kupffer cell surface of the 26-kDa membrane-bound TNF-α isoform [122, 138].

Acquired hepatic dysfunction is a frequent complication of sepsis and trauma [1, 7, 77–79] though the mechanisms are unclear. One postulate is that microbial products and/or the paracrine action of cytokines stimulate formation of ROS which initiate parenchymal injury. Bautista et al. showed that endotoxemia produced increased superoxide anion generation in perfused rat liver [139], while Hewett et al. defined interactions between TNF-α and PMNs in endotoxin-induced liver injury [140]. Most likely sequential Kupffer cell stimulation by microbial products and secondary oxidant stress during ischemia/reperfusion or hypoxia/reoxygenation occur in this setting. However, their sequence dependent effects have not been characterized. In contrast to increases in TNF-α protein during nonbacteremic hypoxia/reoxygenation [137], equivalent hypoxic stress beginning after intraportal *E. coli* bacteremia downregulates TNF-α at a posttranscriptional level, independent of cyclo-oxygenase products or xanthine oxidase-derived ROS.

Metabolic Inactivation of Inflammatory Mediators: Kupffer Cell–Hepatocyte Interactions

Inactivation and detoxification of inflammatory mediators by hepatobiliary elimination is a third critical element of systemic and pulmonary host defense. Circulating and tissue-based concentrations of several classes of mediators implicated in ARDS and multiple organ failure reflect a balance between synthesis and metabolism. Insight into ARDS therefore requires analysis of intrapulmonary cytokine networks as well as hepatic metabolic pathways [141]. TNF-α and eicosanoids, particularly the leukotrienes (LTs), have been studied the most in this regard [61, 72, 110]. However, they are not the only endogenous substances so affected. This variant of the spillover paradigm has been proposed to play a role in chronic, non-ARDS pulmonary expressions of abnormal liver-lung interactions. These include liver disease-induced pulmonary hypertension and the hepatopulmonary syndrome, the triad of liver disease, lung vascular dilations, and abnormal arterial oxygenation [7, 141, 142]. Thus, failure of the damaged liver to inactivate as yet unidentified endogenous compounds can induce acute endothelial injury and inflammation typifying ARDS, chronic pulmonary vasoconstriction, or pulmonary vasodilation in the hepatopulmonary syndrome.

Mechanisms for impaired hepatocytic clearance include changes in the characteristics of Q_L (see above), acute or chronic hepatocytic damage, or a combination of these factors. Accordingly, experimental approaches have targeted one or another of these mechanisms. In most respects the hemodynamic derangements which affect Kupffer cell phagocytic uptake likewise alter hepatobiliary elimination of vasoactive cytokines and eicosanoid lipoxygenation products. A TNF-α:LT axis of inflammation is unmasked by impaired hepatocytic performance. Several lines of evidence suggest that this axis functionally prefers the liver and lungs in ways relevant to ARDS: (a) both TNF-α and LTs undergo extensive hepatic metabolism [144–146]; (b) both are present in increased concentrations in the lower respiratory tract of patients with ARDS or at enhanced risk for ARDS [147, 148]; (c) systemic generation of LTs is enhanced during ARDS [149]; (d) even without sepsis renal excretion of LTs is increased in patients with cirrhosis [150]; (e) TNF-α stimulates production of LTs in vivo [151]; and (f) impairment of hepatocytic performance by the hepatotoxin d-galactosamine increases mortality and lung injury during endotoxemia [152, 153], gram-negative bacteremia, or gram-positive bacteremia, processes mediated by TNF-α but reversible by lipoxygenase inhibition [154, 155].

Changes in flow-dependent performance likewise alters the metabolism of these mediators. Diversion of Qpv after a portacaval shunt selectively decreases Q_L by up to 75% and models reductions in Q_L occurring after sepsis and trauma. In conscious rats with a portacaval shunt we tested the hypothesis that reductions in Q_L amplify shock, lung inflammation, and mortality during endotoxemia by impairing the clearance of endogenously synthesized TNF-α [72]. For equivalent serum endotoxin in shunted and sham-operated animals, mortality, peak circulating TNF-α, neutrophilic alveolitis, and LTB_4 in bronchoalveolar lavage fluid were all increased. Hepatic, cardiac, and cecal microvascular permeability were also greater. Consonant with the evolving understanding that *reciprocal* rather than *unidirectional* interactions occur among mediator classes during endotoxemia, downregulation of LPS-induced TNF-α expression was observed after posttreatment with the lipoxygenase blocker diethylcarbamazine or the 5-lipoxygenase activating protein inhibitor MK-886. This suggested LPS-related production of LTs stimulate TNF-α gene expression, and that such selective reductions in Q_L early during endotoxemia functionally linked the liver and the lungs via a cross-modulating TNF-α:LT axis of inflammation. These findings agree with data obtained during acute liver injury induced by D-galactosamine [154].

Modulation of Inflammatory Responses by Hepatic Acute-Phase Protein Synthesis

The host response to inflammation or tissue injury mediated by the acute-phase response centered in the liver is characterized by a highly coordinated set of immunophysiological reactions ultimately involving the lungs and other organs [156, 157]. That this is a fourth mechanism of systemic and pulmonary defense is supported by two lines of investigation which clarify aspects of the acute-phase response relevant to liver-lung interactions. The first relates to cytokine-mediated responses to acute illness resulting in a reprioritization of the synthesis of acute-phase proteins by the liver [157, 158]. These include α_2-macroglobulin, CRP, serum amyloid A, fibrinogen, and the proteinase inhibitors α_1-antitrypsin and α_1-antichymotrypsin, among others, whose concentrations increase rapidly to exceed those of otherwise phenotypically dominant proteins such as albumin and transferrin [159]. Second, TNF-α, IL-1α, IL-1β, IL-6, and γ-interferon are major regulators of acute-phase gene expression based on the discovery of cell surface receptors, *cis*-acting regulatory elements, and *trans*-acting nuclear proteins that mediate the effects of acute-phase reactants on target genes [157, 160, 161]. The metabolic effects of certain acute-phase reactants support interorgan substrate fluxes while the anti-inflammatory effects of others dampen excessive inflammatory responses. Teleologically these processes might be viewed as facilitating recovery by compensating for impairments in the other three mechanism of host defense.

The acute-phase response orchestrated by the liver alters the interorgan exchange of substrates for protein synthesis and energy production [158, 159], which is balanced by production of nitric oxide which can impair hepatocytic functions [162–164]. The hyperdynamic cardiovascular response to critical illness is paralleled by elevations in VO_2 requirements from increased energy expenditure. This is accompanied by changes in cytokine and counter-regulatory hormone (e.g., epinephrine, cortisol, glucagon) profiles typifying the stress response [165, 167]. The latter process generates alanine from muscle which is a precursor for hepatic gluconeogenesis. At the same time hepatic uptake of circulating amino acids increases which constitute an important oxidative fuel source [168], reflected by ureagensis. Finally, enhanced hepatic pyruvate decarboxylation increases acetyl-coenzyme A entering the Krebs cycle as citrate and drive ATP production [169]. The ATP so produced stabilized hepatic gluconeogenesis, ultimately liberating glucose for host defense functions including white cell energy metabolism. To summarize: (a) Kupffer cell-hepatocyte paracrine interactions after stress balance the stimulatory actions of LPS, TNF-α, IL-1, and especially IL-6 against the inhibitory effects of nitric oxide and other reactive nitrogen intermediates on hepatic protein synthesis; and (b) the metabolic changes initiated by these elements and counterregulatory hormones alters the metabolic milieu directly and by reprioritization of de novo hepatic protein synthesis to acute-phase reactants. These complex interactions have been reviewed [165, 166]. Recent data also support the premise that acute-phase reactants such as α_2-macroglobulin and CRP possess cytoprotective functions.

α-Macroglobulins are glycoproteins (718000 kDa) with proteinase-binding activity, α_2-macroglobulin is secreted by the liver which is abundant in normal human plasma (~2 mg/ml) [170]. Two functions of α_2-macroglobulin are important in cytoprotection: (a) antiproteinase activity and (b) cytokine binding, transport, and/or neutralization [171]. Tissue-damaging proteinases such as collagenase, cathepsin G, and trypsin released by phagocytic cells can be internalized by α_2-macroglobulin via a "molecular trap" mechanism [170]. Clearance of these complexes from the circulation occurs by cellular uptake via α-$_2$-macroglobulin receptors on hepatocytes and macrophages [170, 171]. The pathways by which α_2-macroglobulin exerts regulatory influences on cytokine biology in vivo include the following (from [171]):

- Uptake by low-density lipoprotein receptor-related protein/α_2-macroglobulin receptors
  - Clearance, endocytosis, and lysosomal degradation
  - Delivery of cytokine activity to cells that coexpress low-density lipoprotein receptor-related protein/α_2-macroglobulin receptors

- Antiproteinase activity of α_2-macroglobulin
  - Altered proteolytic activation of latent cytokines
  - Reduced proteolytic degradation of active cytokines

- Proteinase activity (of α_2-macroglobulin-proteinase complexes)
  - Proteolysis of small cytokines in presence of macromolecular antiproteinases

- Cytokine-binding activity
  - Receptor competition
  - Cytokine reservoir or concentration
  - Inactivation

Multiple pro-and anti-inflammatory cytokines are bound by α_2-macroglobulin, including TNF-α, IL-1β, IL-6 TGFβ-1, TGFβ-2, nerve growth factor, and platelet-derived growth factor [172, 173]. Of relevance to determination of the role of specific circulating cytokines in shock and trauma is the transport function of cytokines by α_2-macroglobulin (see above). This carrier process may shield bound but biologically active cytokines from detection systems despite fluxes in interorgan transport and deposition.

CRP is one of a group of structurally related proteins called pentraxins circulating in the blood as molecules composed of five noncovalently associated subunits [157]. Tillett and Francis observed that acute pneumococcal infection was associated with appearance of CRP [174] whose concentration decreases after resolution of acute illness. CRP has multiple activities relevant to pathogenetic events in ARDS and multiple organ failure. CRP levels are elevated in patients with ARDS and these elevations correspond with inhibition of PMN chemotaxis in vitro [175]. Elevated serum CRP concentrations were associated with decreased pulmonary leukosequestration and vascular permeability in a rabbit C5a-induced alveolitis model [176]. Similar inhibition of C5a des arg-induced neutrophil alveolitis was recently confirmed in transgenic mice expressing CRP [177]. The mechanisms(s) by which such inhibition of PMN chemotaxis occurs is incompletely understood. Physical interactions with PMN adhesion molecules, alteration of the PMN cytoskeleton increasing cell stiffness, or changes in the availability/binding of Ca^{2+} to PMNs are being investigated. CRP has other actions relevant to host defense. These include phosphorylcholine-mediated binding to damaged cell membranes [178] and activation of the classical pathway of the complement system with resultant opsonization of particulates and micro-organisms [179].

Conclusion

Here we have outlined a conceptual paradigm concerning the role of the liver in host defense during critical illness. The principal utility of this schema is that it facilitates integration of recent experimental and clinical data from diverse lines of investigation. From this discussion, it is clear that changes in the characteristics of the splanchnic circulation can have significant, but clinically occult consequences on immunoregulatory aspects of hepatic performance. Further mechanistic analyses of this gut–liver axis will be central to an understanding of pivotal organ system interactions in the critically ill.

Acknowledgements. This research was supported in part by NIH grant RO1 GM-43153.

References

1. Helzberg JH, Spiro HM (1986) "LFTs" test more than the liver. JAMA 256:3006–3007
2. Matuschak GM (1992) Multiple systems organ failure: clinical expression, pathogenesis, and therapy. In: Hall JB, Schmidt GA, Wood LDH (eds) Principles of critical care. McGraw-Hill, New York, pp 613–636
3. American College of Chest Physicians/Society of Critical Care Medicine Consensus Conference (1992) Definitions for sepsis and organ failure and guidelines for the use of innovative therapies in sepsis. Crit Care Med 20:864–874
4. Bell RC, Coalson JJ, Smith JD et al (1983) Multiple organ system failure and infection in adult respiratory distress syndrome. Ann Intern Med 99:293–298
5. Montgomery AB, Stager MA, Carrico CJ et al (1985) Causes of mortality in patients with the adult respiratory distress syndrome. Am Rev Respir Dis 132:485–489
6. Pinsky MR, Matuschak GM (1989) Multiple systems organ failure: failure of host defense homeostasis. Crit Care Clin 5:199–220
7. Matuschak GM (1988) Organ interactions in the adult respiratory distress syndrome during sepsis: role of the liver in host defense. Chest 94:400–406
8. Matuschak GM, Rinaldo JE, Pinsky MR et al (1987) Effect of end-stage liver failure on the incidence and resolution of the adult respiratory distress syndrome. J Crit Care 2:162–173
9. Rogers DE (1960) Host mechanisms which act to remove bacteria from the blood. Bacteriol Rev 24:50–66
10. Bradfield JWB (1974) Control of spillover: the importance of Kupffer cell function in clinical medicine. Lancet 2:883–886
11. Saba TM (1970) Physiology and physiopathology of the reticuloendothelial system. Arch Intern Med 126:1031–1035
12. Ballet F (1990) Hepatic circulation: potential for therapeutic intervention. Pharmacol Ther 47:281–328
13. Saba TM (1989) Fibronectin: relevance to phagocytic host response to injury. Circ Shock 29:257–278
14. Mathison JC, Ulevitch RJ (1979) Clearance, tissue distribution and cellular distribution of IV injected LPS in rabbits. J Immunol 123:2133–2143
15. Ruiter DJ, van der Meulen J, Brouwer A et al (1981) Uptake by liver cells of endotoxin following its intravenous injection. Lab Invest 45:38–45
16. Katz MS, Grosfeld JL, Gross K (1984) Impaired bacterial clearance and trapping in obstructive jaundice. Am J Surg 199:14–19

17. Loegering DJ, Blumenstock FA (1985) Depressing hepatic macrophage complement receptor function causes increased susceptibility to endotoxemia and infection. Infect Immun 47:659–664
18. Cuddy BG, Loegering DJ, Blumenstock FA et al (1986) Hepatic macrophage complement receptor clearance function following injury. J Surg Res 40:216–224
19. Decker K (1990) Biologically active products of stimulated liver macrophages (Kupffer cells). Eur J Biochem 29:451–456
20. Fong Y, Marano M, Moldawer L et al (1990) The acute splanchnic and peripheral tissue and metabolic responses to endotoxin in humans. J Clin Invest 85:1896–1904
21. Deitch EA (1993) Gut-liver axis in multiple systems organ failure: role of bacterial translocation and endotoxemia. In: Matuschak GM (ed) Multiple systems organ failure: hepatic regulation of systemic host defense. Dekker, New York, pp 39–72
22. Deitch EA, Sittig K, Li M et al (1990) Obstructive jaundice promotes bacterial translocation from the gut. Am J Surg 159:79–84
23. Nolan JP, Camara DS (1989) Intestinal endotoxins as co-factors in liver injury. Immunol Invest 18:325–337
24. Wilmore DW, Smith RJ, O'Dwyer ST et al (1988) The gut: a central organ after surgical stress. Surgery 104:917–923
25. Marshall JC, Christou NV, Horn H et al (1988) The microbiology of multiple organ failure: the proximal GI tract as an occult reservoir of pathogens. Arch Surg 123:309–315
26. Sori AJ, Rush BF, Lysz TW et al (1988) The gut as a source of sepsis after hemorrhagic shock. Am J Surg 155:187–192
27. Koike K, Moore EE, Moore FA et al (1992) Phospholipase A_2 inhibition decouples lung injury from gut ischemia-reperfusion. Surgery 112:173–180
28. Hechtman HB (1993) Mediators of local and remote injury following gut ischemia. J Vasc Surg 18:134–135
29. Yamaguchi Y, Tamaguchi K, Babb JL et al (1982) In vivo quantitation of the rat liver's ability to eliminate endotoxin from portal vein blood. J Reticuloendothel Soc 32:409–432
30. Fox ES, Thomas P, Broitman SA (1988) Uptake and modification of an ^{125}I-lipopolysaccharide by isolated rat Kupffer cells. Hepatology 8:1550–1554
31. Hewett JA, Roth RA (1993) Hepatic and extrahepatic pathobiology of bacterial lipopolysaccharides. Pharmacol Rev 45:381–411
32. Kaplan JE, Saba TM (1978) Platelet removal from the circulation by the liver and spleen. Am J Physiol 235:H314–H320
33. Schumacker PR, Saba TM (1980) Pulmonary gas exchange abnormalities following intravascular coagulation: reticuloendothelial involvement. Ann Surg 192:95–102
34. Loegering DJ (1983) Intravascular hemolysis and RES phagocytic and host defense functions. Circ Shock 10:383–395
35. Commins LM, Loegering DJ, Gudewicz PW (1990) Effect of erythrocyte and erythrocyte ghost phagocytosis on macrophage phagocytic function and hydrogen peroxide production. Inflammation 14:705–715
36. Donald DE (1983) Splanchnic circulation. In: Shepperd JT, Abbaud FM (eds) Handbook of physiology: the cardiovascular system, section 2, vol 3:peripheral circulation, part 1. American Physiological Society, Bethesda, pp 219–240
37. Hanson KM, Johnson PC (1966) Local control of hepatic arterial and portal venous flow in the dog. Am J Physiol 211:712–717
38. McCluskey RS, Reilly FD (1993) Hepatic microvasculature: dynamic structure and its regulation. Semin Liver Dis13:1–12
39. Gutierrez G, Palizas F, Doglio G et al (1992) Gastric intramucosal pH as a therapeutic index of tissue oxygenation in critically ill patients. Lancet 339:195–199
40. Steffes CP, Dahn MS, Lange MP (1994) Oxygen transport-dependent splanchnic metabolism in the sepsis syndrome. Arch Surg 129:46–52

41. Burton-Opitz R (1911) The vascularity of the liver: the effect of stimulation of single nerves of the hepatic plexus upon the flow in the hepatic artery. Q J Exp Physiol 4:103–116
42. Lautt WW, Legare DJ, D'Almeida MA (1985) Adenosine as a putative regulator of hepatic arterial flow (the buffer response). Am J Physiol 248:H331–H338
43. Greenway CV (1984) Role of splanchnic venous system in overall cardiovascular homeostasis. Fed Proc 42:1678–1684
44. Parks DA, Jacobson ED (1985) Physiology of the splanchnic circulation. Arch Intern Med 145:1278–1281
45. Baker T, Deitch EA, Ma L et al (1988) Hemorrhagic shock promotes the systemic translocation of bacteria from the gut. J Trauma 28:896–906
46. Deitch EA, Bridges W, Ma L et al (1990) Hemorrhagic shock-induced bacterial translocation: the role of neutrophils and hydroxyl radicals. J Trauma 30:942–952
47. Kageshima K, Miyano K, Tanifuji Y et al (1990) Effects of dopamine and dobutamine on hepatic circulation and metabolism in hemorrhagic hypotensive pigs. Masui 39:592–599
48. Nordin A, Makisalo H, Hockersstedt K (1994) Dopamine infusion during resuscitation of experimental hemorrhagic shock. Crit Care Med 22:151–156
49. Caty ML, Guice KS, Oldham KT et al (1990) Evidence for tumor necrosis factor-induced pulmonary microvascular injury after intestinal ischemia-reperfusion. Ann Surg 212:694–700
50. Turnage RH, Guice KS, Oldham KT (1994) Endotoxemia and remote organ injury following intestinal reperfusion. J Surg Res 56:571–578
51. Bark T, Katouli M, Ljungqvist S et al (1993) Bacterial translocation after non-lethal hemorrhage in the rat. Circ Shock 41:60–65
52. Deitch EA, Berg RD (1987) Endotoxin but not malnutrition promotes bacterial translocation from the gut. J Trauma 27:161–166
53. Koziol JM, Rush BF Jr, Smith SM et al (1988) Occurrence of bacteremia during and after hemorrhagic shock. J Trauma 28:10–16
54. Nelson DP, Samsel RW, Wood LDH et al (1988) Pathological supply dependence of systemic and intestinal O_2 uptake during endotoxemia. J Appl Physiol 64:2410–2419
55. Tracey KJ, Lowry SF, Fahey TJ III et al (1987) Cachectin/tumor necrosis factor induces lethal shock and stress hormone responsiveness in the dog. Surgery 164:415–422
56. Tracey KJ, Fong Y, Hesse DG et al (1987) Anti-cachectin/TNF monoclonal antibodies prevent septic shock during lethal bacteremia. Nature 330:662–664
57. Dinarello CA (1991) The proinflammatory cytokines interleukin-1 and tumor necrosis factor and treatment of the septic shock syndrome. J Infect Dis 163:1177–1184
58. Van Zee KJ, LaForge LE, Fischer E et al (1991) IL-8 in septic shock, endotoxemia, and after IL-1 administration. J Immunol 146:3478–3482
59. Hagmann WE, Denzlinger C, Keppler D (1984) Role of peptide leukotrienes and their hepatobiliary elimination in endotoxin action. Circ Shock 14:233–238
60. Robbins RA, Russ WA, Rasmussen JK (1987) Activation of the complement system in the adult respiratory distress syndrome. Am Rev Respir Dis 135:651–658
61. Jaeschke H, Raftery MJ, Justesen U et al (1992) Serum complement mediates endotoxin-induced cysteinyl leukotriene formation in rats in vivo. Am J Physiol (Gastrointest Liver Physiol) 263:G947–G952
62. Bone RC (1991) The pathophysiology of sepsis. Ann Intern Med 115:457–469
63. Giroir BP (1993) Mediators of septic shock: new approaches for interrupting the endogenous inflammatory cascade. Crit Care Med 21:780–789
64. Natanson C, Hoffman WD, Suffredini AF et al (1994) Selected treatment strategies for septic shock based on proposed mechanisms of pathogenesis. Ann Intern Med 120:771–783
65. Ciancio MJ, Hunt J, Jones SB et al (1991) Comparative and interactive in vivo effects of tumor necrosis factor a and endotoxin. Circ Shock 33:108–120

66. Prytz H, Holst-Christenson J, Korner B et al (1976) Portal venous and systemic endotoxaemia in patients without liver disease and systemic endotoxaemia in patients with cirrhosis. Scand J Gastroenterol 11:857–863
67. Lumsden AB, Henderson JM, Kutner MN (1988) Endotoxin levels measured by a chromogenic assay in portal, hepatic and peripheral venous blood in patients with cirrhosis. Hepatology 8:232–236
68. Wilkinson SP, Moodie H, Stamatakis JD et al (1976) Endotoxaemia and renal failure in cirrhosis and obstructive jaundice. Br Med J 2:1415–1418
69. Goldfarb G, Nouel O, Poynard T et al (1983) Efficiency of respiratory assistance in cirrhotic patients with liver failure. Intensive Care Med 9:271–273
70. Matuschak GM, Shaw BW Jr (1987) Adult respiratory distress syndrome associated with acute liver allograft rejection: resolution following hepatic retransplantation. Crit Care Med 15:878–881
71. Decamp MM, Warner AE, Molina RM et al (1992) Hepatic versus pulmonary uptake of particles injected into the portal circulation in sheep. Am Rev Respir Dis 146:224–231
72. Matuschak GM, Mattingly M, Tredway TL et al (1994) Liver-lung interactions during E. coli endotoxemia: TNF-α:leukotriene axis. Am J Respir Crit Care Med 149:41–49
73. Halvorsen L, Roth R, Gunther RA et al (1993) Liver hemodynamics during portal venous endotoxemia in swine. Circ Shock 41:166–175
74. Raven HA, Fine J (1962) Biological implications of intestinal endotoxin. Fed Proc 21:65–68
75. Nolan JP, Leibowitz AL (1978) Endotoxin and the liver. III. Modification of acute carbon tetrachloride injury by polymyxin B – an antiendotoxin. Gastroenterology 75:445–449
76. Fox ES, Broitman, Thomas P (1990) Bacterial endotoxins and the liver. Lab Invest 63:733–741
77. Franson TR, Hierholzer WJ Jr, LaBreque DR (1985) Frequency and characteristics of hyperbilirubinemia associated with bacteremia. Rev Infect Dis 7:1–9
78. Gimson AES (1987) Hepatic dysfunction during bacterial sepsis. Intensive Care Med 13:162–166
79. von Allmen D, Hasselgren P-O, Higashiguchi T et al (1992) Individual regulation of different hepatocellular functions during sepsis. Metabolism 41:961–965
80. Soons PA, De Boer A, Cohen AF et al (1991) Assessment of hepatic blood flow in healthy subjects by continuous infusion of indocyanine green. Br J Clin Pharmacol 32:697–704
81. Suren A, Bauer FE, Rosenkranz B et al (1991) Effect of pentoxifylline on liver plasma flow in normal man. Eur J Clin Pharmacol 41:233–237
82. Leen E, Angerson WJ, Warren HW et al (1993) Duplex/colour Doppler sonography: measurement of changes in hepatic arterial hemodynamics following intra-arterial angiotensin II infusion. Br J Cancer 67:1381–1384
83. Dahn MS (1994) Hepatic dysfunction in the critically ill and injured. Intensive Care World 11:9–14
84. Perlik F, Janku I, Jedlicka J (1992) Functional and non-functional liver blood flow in patients with liver cirrhosis measured by indocyanine green. Meth Find Exp Clin Pharmacol 14:459–464
85. Dahn MS, Lange P, Wilson RF et al (1990) Hepatic blood flow and splanchnic oxygen consumption measurements in clinical sepsis. Surgery 107:295–301
86. Wang P, Zheng FBA, Chaudry IH (1991) Increase in hepatic blood flow during early sepsis is due to increased portal flow. Am J Physiol 261 (Regulatory Integrative Comp Physiol 30):R1507–R1512
87. Wang P, Zheng FBA, Ayala A et al (1992) Hepatocellular dysfunction persists during early sepsis despite increase volume of crystalloid resuscitation. J Trauma 32:389–397

88. Wang P, Ba ZF, Burkhardt J et al (1992) Measurement of hepatic blood flow after severe hemorrhage: lack of restoration despite adequate resuscitation. Am J Physiol 262 (Gastrointest Liver Physiol 6): G92–G98
89. Marzi I, Bauer C, Hower R et al (1993) Leukocyte-endothelial interactions in the liver after hemorrhagic shock in the rat. Circ Shock 40:105–114
90. Jaeschke H (1991) Vascular oxidant stress and hepatic ischemia/reperfusion injury. Free Radic Res Commun 12–13:737–743
91. Johnson EE, Hedley-Whyte J (1977) End-expiratory pressure ventilation and sulfobromophthalein sodium excretion in dogs. J Appl Physiol 43:714–720
92. Rossaint R, Krebs M, Forther J et al (1993) Inferior vena caval pressure increase contributes to sodium and water retention during PEEP in awake dogs. J Appl Physiol 75:2484–2492
93. Matuschak GM, Pinsky MR (1989) Effects of positive-pressure ventilatory frequency on hepatic blood flow and performance. J Crit Care 4:153–165
94. Johnson EE, Hedley-Whyte J (1972) Continuous positive-pressure ventilation and portal flow in dogs with pulmonary edema. J Appl Physiol 33:385–389
95. Winso O, Biber B, Gustavsson B et al (1986) Portal blood flow in man during graded positive end-expiratory pressure ventilation. Intensive Care Med 12:80–85
96. Bennett TD, Rothe C (1981) Hepatic capacitance responses to changes in flow and hepatic venous pressure in dogs. Am J Physiol 239 (Heart Circ Physiol 9): H474–H481
97. Richard C, Berdeaux A, Delion F et al (1986) Effect of mechanical ventilation on hepatic drug pharmacokinetics. Chest 90:837–841
98. Matuschak GM, Pinsky MR, Rogers RM (1987) Effects of positive end-expiratory pressure on hepatic blood flow and performance. J Appl Physiol 62:1377–1383
99. Purcell PN, Branson RD, Hurst JM et al (1992) Gut feeding and hepatic hemodynamics during PEEP ventilation for acute lung injury. J Surg Res 53:335–341
100. Matuschak GM, Munoz C, Epperly NA et al (1994) TNF-α and IL-6 expression after intraportal candidemia vs. E. coli or S. aureus bacteremia. Am J Physiol (Regul Integrative Comp Physiol) 267: R446-R454
101. Matuschak GM, Lechner AJ (1993) Hepatic regulation of systemic host defense and its derangement in multiple systems organ dysfunction and failure. In: Matuschak GM (ed) Multiple systems organ failure: hepatic regulation of systemic host defense. Dekker, New York, pp 1–38
102. Matsumoto K, Tajima H, Hamanque M et al (1992) Identification and characterization of "injurin", an inducer of expression of the gene for hepatocyte growth factor. Proc Natl Acad Sci USA 89:3800–3804
103. Shreck R, Rieber P, Baeurle PA (1991) Reactive oxygen intermediates as apparently widely used messengers in the activation of the NF-κB transcription factor and HIV-1. EMBO J 10:2247–2258
104. Zuckerman SH, Evens GF, Guthrie L (1991) Transcriptional and post-transcriptional mechanisms involved in the differential expression of LPS-induced IL-1 and TNF mRNA. Immunology73:460–465
105. Han J, Huez G, Beutler B (1991) Interactive effects of the tumor necrosis factor promoter and 3′-untranslated regions. J Immunol 146:1843–1848
106. Mohler KM, Torrance DS, Smith CA et al (1993) Soluble tumor necrosis factor (TNF) receptors are effective therapeutic agents in lethal endotoxemia and function simultaneously as both TNF carriers and TNF antagonists. J Immunol 151:1548–1561
107. Steinkasserer A, Estaller C, Weiss EH et al (1992) Human interleukin-1 receptor antagonist is expressed in liver. FEBS Lett 310:60–62
108. Okusawa S, Yancey KB, van der Meer JWM et al (1988) C5a stimulates secretion of tumor necrosis factor from human mononuclear cells in vitro. J Exp Med 168:443–448

109. Pittner RA, Spitzer JA (1992) Endotoxin and TNFα directly stimulate nitric oxide formation in cultured rat hepatocytes from chronically endotoxemic rats. Biochem Biophys Res Commun 185:430–435
110. Karck U, Peters T, Decker K (1988) The release of tumor necrosis factor from endotoxin-stimulated rat Kupffer cells is regulated by prostaglandin E_2 and dexamethasone. J Hepatol 7:352–361
111. Tran-Thi, T-H, Weinhold L, Weinstock C et al (1993) Production of tumor necrosis factor-α, interleukin-1 and interleukin-6 in the perfused rat liver. Eur Cytokine Netw 4:363–370
112. Khoruts A, Stahnke L, McClain CJ et al (1991) Circulating tumor necrosis factor, interleukin-1 and interleukin-6 concentrations in chronic alcoholic patients. Hepatology 13:267–276
113. Deviere J, Content J, Denys C et al (1990) Excessive in vitro bacterial lipopolysaccharide-induced production of monokines in cirrhosis. Hepatology 11:628–634
114. Torre D, Zeroli C, Giola M et al (1994) Serum levels of interleukin-1α, interleukin-1β, interleukin-6, and tumor necrosis factor in patients with acute viral hepatitis. Clin Infect Dis 18:194–198
115. Muto Y, Nouri-Aria KT, Meager A et al (1988) Enhanced tumour necrosis factor and interleukin-1 in fulminant hepatic failure. Lancet 2:72–74
116. Tilg H, Wilmer A, Vogel W et al (1992) Serum levels of cytokines in chronic liver diseases. Gastroenterology103:264–274
117. Busam KJ, Homfeld A, Zawatsky R et al (1990) Virus vs endotoxin-induced activation of liver macrophages. Eur J Biochem 191:577–582
118. Pinsky MR, Vincent J-L, Deviere J et al (1993) Serum cytokine levels in human septic shock: relation to multiple-system organ failure and mortality. Chest 103:565–575
119. Bodey GP (1993) Hematogenous and major organ candidiasis. In: Bodey GP (ed) Candidiasis: pathogenesis, diagnosis and treatment. Raven, New York, pp 279–329
120. Tang E, Tang G, Berne TV (1993) Prognostic indicators in fungemia of the surgical patient. Arch Surg 128:759–763
121. Gianotti L, Alexander JW, Fukushima R et al (1993) Translocation of Candida albicans is related to the blood flow of individual intestinal villi. Circ Shock 40:250–257
122. Matuschak GM, Klein CA, Tredway TL et al (1993) TNF-α and cyclooxygenase metabolites do not modulate C. albicans septic shock with disseminated candidiasis. J Appl Physiol 74:2432–2442
123. Calderone RA, Braun PC (1991) Adherence and receptor relationships of Candida albicans. Microbiol Rev 55:1–20
124. Olynyk J, Matuschak GM, Lechner AJ et al (1994) Differential production of tumor necrosis factor-α by rat Kupffer cells in vitro after phagocytosis of E. coli and C. albicans. Am J Physiol (Gastrointest Liver Physiol) 267:G213–G219
125. Lechner AJ, Tredway TL, Brink D et al (1992) Differential systemic and intrapulmonary TNF-α production in Candida sepsis during immunosuppression. Am J Physiol (Lung Cell Mol Physiol 7) 263:L526–L535
126. Dahn MS, Wilson RF, Lange MP et al (1990) Hepatic parenchymal oxygen tension following injury and sepsis. Arch Surg 125:441–443
127. Schlichtig R, Klions HA, Kramer DJ et al (1992) Hepatic dysoxia commences during O_2 supply dependence. J Appl Physiol 72:1499–1505
128. Fitch KA, Rink RD (1983) Hepatic oxygen supply during early and late sepsis in the rat. Circ Shock 10:51–59
129. Engerson TD, McKelvey TG, Rhyne DB et al (1987) Conversion of xanthine dehydrogenase to oxidase in ischemic rat tissues. J Clin Invest 79:1564–1570
130. Gonzales-Flecha B, Cutrin JC, Boveris A (1993) Time course and mechanisms of oxidative stress and tissue damage in rat liver subjected to in vivo ischemia-reperfusion. J Clin Invest 91:456–464
131. Chaudri G, Clark IA (1989) Reactive oxygen species facilitate the in vitro and in vivo lipopolysaccharide-induced release of tumor necrosis factor. J Immunol 143:1290–1294

132. Colletti LM, Remick DG, Burtch S et al (1990) Role of tumor necrosis factor-α in the pathophysiologic alterations after hepatic ischemia/reperfusion injury in the rat. J Clin Invest 85:1936–1943
133. Jaeschke H, Farhood A (1991) Neutrophil and Kupffer cell-induced oxidant stress and ischemia-reperfusion injury in rat liver. Am J Physiol (Gastrointest Liver Physiol) 260:G355–G362
134. Jaeschke H, Farhood A, Bautista AP et al (1993) Functional inactivation of neutrophils with a Mac-1 (CD11b/CD18) monoclonal antibody protects against ischemia-reperfusion injury in rat liver. Hepatology 17:915–923
135. Thorne J, Blomquist S, Elmer O et al (1989) Polymorphonuclear leukocyte sequestration in the lungs and liver following soft-tissue trauma: an in vivo study. J Trauma 29:451–456
136. Ayala A, Perrin MM, Wang P et al (1992) Hemorrhage induces enhanced Kupffer cell cytotoxicity while decreasing peritoneal or splenic macrophage capacity: involvement of cell-associated TNF and reactive nitrogen. J Immunol 147:4147–4152
137. Wibbenmeyer LW, Lechner AJ, Munoz CF, Matuschak GM (1995) Downregulation of E. coli-induced TNF-α expression in perfused liver by hypoxia-reoxygenation. Am J Physiol 268 (Gastrointest Liver Physiol 31):G311–G319
138. Keogh C, Fong Y, Mariano MA et al (1990) Identification of a novel tumor necrosis factor-α/cachectin from the livers of burned and infected rats. Arch Surg 125:79–85
139. Bautista AP, Spitzer JJ (1990) Superoxide anion generation by in situ perfused rat liver: effect of in vivo endotoxin. Am J Physiol (Gastrointest Liver Physiol) 259:G907–G912
140. Hewett JA, Jean PA, Kunkel SL et al (1993) Relationship between tumor necrosis factor-α and neutrophils in endotoxin-induced liver injury. Am J Physiol (Gastrointest Liver Physiol) 265:G1011–G1015
141. Bernard GR, Artigas A, Brigham KL et al (1994) The American-European Consensus Conference on ARDS. Am J Respir Crit Care Med 149:818–824
142. Krowka MJ (1993) Clinical management of the hepatopulmonary syndrome. Semin Liver Dis 13:414–422
143. Chang S-W, Ohara N (1992) Pulmonary circulatory dysfunction in rats with biliary cirrhosis: an animal model of the hepatopulmonary syndrome. Am Rev Respir Dis 145:798–805
144. Beutler BA, Milsark IW, Cerami A (1985) Cachectin/tumor necrosis factor: production, distribution, and metabolic fate in vivo. J Immunol 135:3972–3977
145. Denzlinger C, Rapp S, Hagmann W et al (1985) Leukotrienes as mediators in tissue trauma. Science 230:330–332
146. Denzlinger C, Gohlmann A, Scheuber PH et al (1986)Metabolism and analysis of cysteinyl leukotrienes in the monkey. J Biol Chem 25:15601–15606
147. Suter PM, Suter S, Girardin E et al (1992) High bronchoalveolar levels of tumor necrosis factor and its inhibitors, interleukin-1, interferon, and elastase, in patients with adult respiratory distress syndrome after trauma, shock, or sepsis. Am Rev Respir Dis 145:1016–1022
148. Stephenson AH, Lonigro AJ Hyers TM et al (1985) Increased concentrations of leukotrienes in bronchoalveolar lavage fluid of patients with ARDS or at risk for ARDS. Am Rev Respir Dis 138:714–719
149. Bernard GR, Korley V, Swindell B et al (1991) Persistent generation of peptide leukotrienes in patients with the adult respiratory distress syndrome. Am Rev Respir Dis 144:263–267
150. Huber M, Kastner S, Scholmerich J et al (1989) Cysteinyl leukotriene analysis in human urine: enhanced excretion in patients with liver cirrhosis and hepatorenal syndrome. Eur J Clin Invest 19:53–60
151. Huber M, Beutler B, Keppler D (1988) Tumor necrosis factor-α stimulates leukotriene production in vivo. Eur J Immunol 18:2085–2088

152. Lehmann V, Freudenberg MA, Galanos C (1987) Lethal toxicity of lipopolysaccharide and tumor necrosis factor in normal and D-galactosamine-treated mice. J Exp Med 165:657–663
153. Freudenberg MA, Galanos C (1991) Tumor necrosis factor alpha mediates lethal activity of killed gram-negative and gram-positive bacteria in D-galactosamine-treated mice. Infect Immun 59:2110–2115
154. Matuschak GM, Pinsky MR, Klein EC et al (1990) Effects of D-galactosamine-induced acute liver injury on mortality and pulmonary responses to E. coli lipopolysaccharide: modulation by arachidonic acid metabolites. Am Rev Respir Dis 141:1296–1306
155. Hagmann W (1993) The liver in leukotriene action, metabolism, and production: physiologic and pathophysiologic conditions. In: Matuschak GM (ed) Multiple systems organ failure: hepatic regulation of systemic host defense. Dekker, New York, pp 165–191
156. Perlmutter DH, Dinarello CA, Punsai PI et al (1986) Cachectin/tumor necrosis factor regulates hepatic acute-phase gene expression. J Clin Invest 78:1349–1354
157. Perlmutter DH (1993) Cytokines and the hepatic acute-phase response. In: Matuschak GM (ed) Multiple systems organ failure: hepatic regulation of systemic host defense. Dekker, New York, pp 229–265
158. Scanga G, Siegal JH, Brown G et al (1985) Reprioritization of hepatic plasma protein release in trauma and sepsis. Arch Surg 120:187–199
159. Cereghini S, Raymonjean M, Carranca AG et al (1987) Factors involved in control of tissue-specific expression of albumin gene. Cell 50:627–638
160. Castell JV, Andus T, Kunz D et al (1989) Interleukin-6: the major regulator of acute-phase protein synthesis in man and rat. Ann NY Acad Sci 169:333–338
161. Korber B, Merod N, Hood L et al (1988) Regulation of gene expression by interferons: control of H-2 promoter responses. Science 239:1302–1306
162. Billiar TR, Curran RD, Stuehr DJ et al (1990) Kupffer cell:hepatocyte cocultures produce nitric oxide in response to bacterial endotoxin. J Surg Res 48:349–355
163. Curran RD, Billiar TR, Stuehr DJ et al (1990) Multiple cytokines are required to induce hepatocyte nitric oxide production and inhibit total protein synthesis. Ann Surg 212:462–471
164. Peters T, Karck U, Decker K (1990) Interdependence of tumor necrosis factor, prostaglandin E_2, and protein synthesis in lipopolysaccharide-exposed rat Kupffer cells. Eur J Biochem 191:583–589
165. Goldstein SA, Elwyn DH (1989) The effects of injury and sepsis on fuel utilization. Annu Rev Nutr 9:445–473
166. Lowry SF (1992) Modulating the metabolic response to injury and infection. Proc Nutr Soc 51:267–277
167. Gelfand RA, Matthews DE, Bier DM et al (1984) Role of counter-regulatory hormones in the catabolic response to stress. J Clin Invest 200:2238–2248
168. Pittiruti M, Siegel JH, Sganga G et al (1989) Determinants of urea-production in sepsis: muscle catabolism, TPN, and hepatic clearance of amino acids. Arch Surg 124:362–372
169. Vary TC, Siegel JH, Nakatani T et al (1986) Effect of sepsis on activity of pyruvate dehydrogenase complex in skeletal muscle and liver. Am J Physiol 13:E634–E640
170. Sottrup-Jensen L (1989) α-Macroglobulins: structure, shape, and mechanism of proteinase complex formation. J Biol Chem 264:11539–11542
171. Wollenberg GK, LaMarre J, Hayes MA (1993) Cytokine interactions with α_2–macroglobulin. In: Matuschak GM (ed) Multiple systems organ failure: hepatic regulation of systemic host defense. Dekker, New York, pp 293–320
172. Wollenberg GK, LaMarre J, Rosendal S et al (1991) Binding of tumor necrosis factor alpha to activated forms of human plasma alpha$_2$ macroglobulin. Am J Pathol 138:265–272

173. Borth W, Luger TA (1989) Identification of α_2–macroglobulin as a cytokine binding plasma protein. Binding of interleukin-1β to "F" α_2–macroglobulin. J Biol Chem 264:5818–5825
174. Tillet WS, Francis T Jr (1930) Serologic reactions in pneumonia with a non-protein somatic fraction of pneumococcus. J Exp Med 52:561–571
175. Kew RR, Hyers TM, Webster RO (1990) Human C-reactive protein inhibits neutrophil chemotaxis in vitro: possible implications for the adult respiratory distress syndrome. J Lab Clin Med 115:339–345
176. Heuertz RM, Piquette CA, Webster RO (1993) Rabbits with elevated serum C-reactive protein exhibit diminished neutrophil infiltration and vascular permeability in C5a-induced alveolitis. Am J Pathol 142:319–328
177. Heuertz RM, Xia D, Samols D et al (1994) Inhibition of C5a des arg-induced neutrophil alveolitis in transgenic mice expressing C-reactive protein. Am J Physiol (Lung Cell Mol Physiol) 266:L649–L654
178. Siegel J, Rent R, Gerwurz H (1975) Interactions of C-reactive protein with the complement system. J Exp Med 142:709–721
179. Geller DA, Kispert PH, Su GL et al (1993) Induction of hepatocyte lipopolysaccharide binding protein in models of sepsis and the acute-phase response. Arch Surg 128:22–28

Lung-Liver Interface

R. Rodriguez-Roisin

One of the most provocative examples of lung-liver interface is the hepatopulmonary syndrome (HPS), a syndrome defined as a clinical triad of chronic liver disease, most commonly all types of hepatic cirrhosis and chronic active hepatitis, gas exchange abnormalities resulting ultimately in severe arterial hypoxemia (very low arterial oxygen tension, $PaO_2 < 60$ mmHg), and intrapulmonary vascular dilatations, in the absence of intrinsic primary cardiopulmonary disease [1]. Additional features may include platypnea (dyspnea induced by the upright position) and orthodeoxia (arterial hypoxemia enhanced by the upright position; decrease in $PaO_2 \geq 10\%$ from supine) with or without shortness of breath, low transfer factor (diffusing capacity for carbon monoxide), and hemodynamic abnormalities (systemic and pulmonary vasodilatation); yet the prevalence of these latter features varies among patients.

From a structural viewpoint the most representative microscopic abnormality of HPS is a normal terminal bronchiole surrounded by a neighboring pulmonary artery extremely enlarged, within an otherwise intact pulmonary architecture. The presence of widespread vasodilatation of the pulmonary vasculature at the precapillary level is the most prominent pathological marker [2]. Although larger arteriovenous communications and pleural spider nevi may be present, they are usually uncommon. From a pathophysiological point of view this structural derangement of the pulmonary circulation at both the precapillary and capillary levels near to the gas exchange zone interferes with an adequate matching of ventilation and perfusion within the alveoli. This is sufficient to allow mixed venous blood to pass directly into the pulmonary veins, hence jeopardizing the mechanisms of normal arterial oxygenation.

In patients with echocardiographic evidence of intrapulmonary vascular dilatations severe arterial hypoxemia in nearly 60% and a low gas transfer factor in 83% of the cases constitute the most common routine gas exchange abnormalities [3]. In addition, both a reduced carbon dioxide arterial tension ($PaCO_2$) and an increased alveolar-arterial O_2 pressure difference reflect preferentially the continuous presence of hyperventilation, an extremely common finding in candidates for liver transplantation, possibly a population with a high prevalence of HPS [4]. The role played by an increased ventilation and particularly by a hyperkinetic hemodynamic circulation, namely an inordinately high cardiac output (see below), is crucial to the understanding of the pathophysiology of HPS.

It should be emphasized that the values of PaO_2 and $PaCO_2$ are the end-point variables of the complex interplay between the intrapulmonary factors (ventilation-perfusion – $\dot{V}_A/\dot{Q}$ – inequalities, increased intrapulmonary shunt, and diffusion alveolar to endcapillary disequilibrium to O_2) and the extrapulmonary components (inspired O_2 tension, overall ventilation, cardiac output, and O_2 consumption) that modulate pulmonary gas exchange. Any increase in the first three extrapulmonary factors and/or a decrease in O_2 uptake increases or enhances PaO_2, others things being equal, and vice versa [5].

What are the specifical causes of arterial hypoxemia in the HPS? This question has been addressed by means of the multiple inert gas elimination technique, a tool which allows to direct estimation of $\dot{V}_A/\dot{Q}$ relationships.

Studies with the multiple inert gas elimination technique have shown $\dot{V}_A/\dot{Q}$ mismatching as the most common mechanism of gas exchange disturbances even in the absence of marked decrements in PaO_2 [1, 6]. The finding of $\dot{V}_A/\dot{Q}$ imbalance in the absence of hypoxemia may be explained by the beneficial effects of both increased ventilation and cardiac output in maximizing arterial oxygenation despite $\dot{V}_A/\dot{Q}$ alterations. In addition, patients with HPS and severe hypoxemia also have a substantial increase in intrapulmonary shunt, and a few studies have found diffusion limitation to O_2. The constant finding of $\dot{V}_A/\dot{Q}$ inequalities is commensurate with the presence of alveoli with low $\dot{V}_A/\dot{Q}$ ratios, in which ventilation is maintained but perfusion is greatly increased. Similarly the coexistence of shunt is in keeping with the finding of true arteriovenous communications in the most severe cases of HPS. Although alveolar to endcapillary diffusion limitation to O_2 may coexist, its influence may be negligible if hypoxemia is not severe enough. A "diffusion-perfusion" defect [7] may cause inadequate diffusion of O_2 to the center of the enlarged capillaries, a condition which is enhanced by a very short transit time of the red blood cell in the pulmonary capillary network due to an inordinately high cardiac output.

The role of increased cardiac output in influencing the pulmonary gas exchange status of HPS is of paramount importance to the understanding of many of the clinical findings of HPS. One of the most representative examples of the former can be demonstrated by means of the hemodynamic behavior of patients with mild HPS during exercise. During submaximal progressive exercise at approximately 60% of the maximal tolerated work load O_2 uptake in patients with HPS was similar to that developed by healthy volunteers (approximately between 1200 and 1300 ml/min; Fig. 1) [8]. Nonetheless, patients with HPS showed cardiac output values that were close to those of normal individuals during almost twice the work load (120 W). These data suggest that the hyperkinetic status generally shown in patients with HPS at rest is still preserved during exercise. Of note is that the driving pulmonary pressure (pulmonary artery pressure minus wedge pressure) during exercise remained within normal limits. The lack of changes in pulmona-

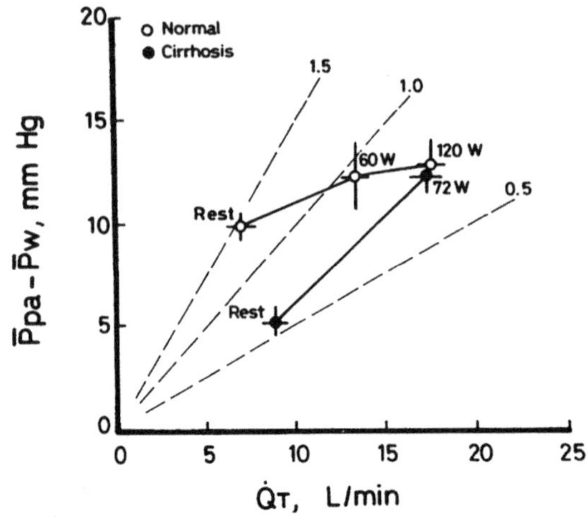

Fig. 1. Pulmonary driving pressure (*Ppa–Pw*) versus cardiac output (Q_T) corresponding to patients with liver cirrhosis associating HPS (*closed circles*) and also to healthy controls (*open circles*). *Dashed tracings irradiating from the origin* correspond to isopulmonary vascular resistance lines. Note that compared to healthy individuals patients with HPS have a hyperkinetic status, characterized by higher cardiac output values and both lower pulmonary driving pressures and vascular resistance. For further explanation, see text. (From [8])

ry vascular resistance during exercise is at variance with the behavior usually shown in normal individuals who are able to recruit and distend more pulmonary capillaries and induce a reduction of pulmonary vascular resistance. This is akin to the concept that patients with HPS develop an almost completely dilated and recruited pulmonary vascular bed even at rest. Thus any further increase in cardiac output necessarily results in higher pulmonary vascular pressures without decreasing pulmonary vascular resistance. However, these hemodynamic changes did not induce changes in pulmonary gas exchange efficiency per se since $\dot{V}_A/\dot{Q}$ mismatch remained essentially unaltered, and no diffusion limitation for the O_2 transfer ensued.

Another cardinal feature of HPS is the behavior of the pulmonary vascular tone. We studied the hypoxic vascular response in patients with liver cirrhosis, half of them probably having HPS [6]. Overall, while breathing a hypoxic mixture pulmonary vascular resistance increased, whereas $\dot{V}_A/\dot{Q}$ inequalities remained unchanged. This suggests that there was still a resting pulmonary vascular reactivity. Interestingly, when patients were divided according to the presence or absence of spiders, patients with spiders, who also showed greater liver dysfunction and were considered carriers of an HPS, had a lower hypoxic vascular response. Furthermore, patients with spiders developed more $\dot{V}_A/\dot{Q}$ mismatch, as shown by greater increases in intrapulmonary shunt and in the dispersion of pulmonary blood flow, an index sensitive to below-normal and low $\dot{V}_A/\dot{Q}$ ratios, upon breathing 100% O_2. In addition, patients with HPS who had greater liver dysfunction had also more

$\dot{V}_A/\dot{Q}$ inequalities irrespective of the FiO_2 breathed. These data are consistent with enhancement of the development of alveolar units with low $\dot{V}_A/\dot{Q}$ ratios by inadequate pulmonary vascular tone. These findings are in keeping with the concept that a low pulmonary vascular tone aggravates pulmonary gas exchange whereas enhancement of pulmonary vascular tone ameliorates it [9].

Enhancement of arterial hypoxemia by the upright position, also known as orthodeoxia, a remarkable finding seen in patients with HPS, has been shown to be associated with higher arterial to venous O_2 content difference, lower mixed venous PO_2, and possibly lower cardiac output without changes in intrapulmonary shunt than when lying supine [10]. Conceivably, gravitational blood flow differences of the pulmonary circulation are maximized in the upright position. This may induce further deterioration in $\dot{V}_A/\dot{Q}$ mismatch, possibly reinforced by a lower pulmonary vascular tone. However, this hypothesis needs to be explored by measuring more directly the amount of $\dot{V}_A/\dot{Q}$ mismatch in each position.

Among the many medical therapeutic options tested in patients with HPS, such as the administration of somastotatin analogue, almitrine bismesylate, or prostaglandins [11], one of the most promising approaches from both a hemodynamic and a gas exchange viewpoint, was the administration of propranolol. The treatment with this beta-blocker has been advocated in patients with advanced liver cirrhosis and portal hypertension because by reducing heart rate and cardiac output [12] it lowers portal pressure and gastroesophageal collateral blood flow.

We treated patients with HPS with propranolol at a loading dose of 0.1 mg/kg followed by a continuous intravenous perfusion of 2 mg/h [13]. Propranolol lowered cardiac output, thereby decreasing intrapulmonary shunt, and improved the $\dot{V}_A/\dot{Q}$ distributions. However, propranolol failed to improve the distribution of pulmonary blood flow in relation to that of alveolar ventilation. Interestingly, PaO_2 remained unchanged following the administration of the drug because the detrimental effect of the reduction in cardiac output on PaO_2 was offset by two beneficial effects: on the one hand, as alluded to, intrapulmonary shunt decreased; on the other, the efficiency of the lung as a gas exchanger was improved, as shown by an amelioration in both $\dot{V}_A/\dot{Q}$ ratio distributions. It is nevertheless of interest to point out some aspects related to peripheral gas exchange. Because cardiac output decreased after propranolol, O_2 delivery (namely, arterial O_2 content times cardiac output) also fell. However, O_2 consumption remained unaltered because the extraction ratio of O_2, that is arterial to venous O_2 content difference divided by arterial O_2 content, increased substantially. This effect of propranolol in patients with HPS is similar to that in healthy individuals. Thus, under normal conditions O_2 uptake is determined by tissue metabolic activity and is independent of O_2 delivery. When the latter parameter is reduced, O_2 uptake remains constant because the extraction of O_2 by the tissues remains unchanged [14]. It has been suggested that patients with severe liver disease may present an abnormal capacity to regulate tissue O_2 extraction which would then render O_2 consumption pathologically dependent on the delivery of O_2 [15]. Our data clearly show that in clinically stable patients with HPS any reduction in O_2 delivery is fully compensated by an adequate increase in the extraction of O_2 by the tissues. Moreover, they do not support the contention that these patients have subclinical tissue hypoxia or an impaired ability to regulate tissue O_2 extraction. However, even in normal individuals O_2 uptake becomes linearly dependent on O_2 supply below a critical value of O_2 delivery [14]. In anesthesized patients undergoing coronary surgery 8.2 ml min^{-1} kg^{-1} is considered to be the figure corresponding to this critical O_2 delivery value [16]. None of the patients studied by us was below this critical value. Conceivably, this was due to the fact that these patients were essentially normoxemic or slightly hypoxemic and had higher than normal O_2 delivery values because of their hyperkinetic status. It is possible that under more critical conditions in patients with HPS, such as anemia, more arterial hypoxemia or more liver failure, that is with a lower initial O_2 delivery, propranolol decreases it to below a critical level that causes tissue hypoxia. From a clinical viewpoint it can therefore be recommended to monitor arterial O_2 content routinely in patients with hepatic cirrhosis to be considered for propranolol therapy. Further, those patients who have a low O_2 content at baseline should be closely controlled during treatment if propranolol is eventually administered.

Acknowledgements. This research was supported in part by the Comissionat per a Universitats i Recerca de la Generalitat de Catalunya (GRQ94-9103).

References

1. Rodriguez-Roisin R, Agustí AGN, Roca J (1992) The hepatopulmonary syndrome: new name, old complexities. Thorax 47:897–902
2. Berthelot P, Walker JG, Sherlock S, Reid L (1966) Arterial changes in the lungs in cirrhosis of the liver-lung spider nevi. N Engl J Med 274:291–298
3. Krowka MJ, Dickson ER, Cortese DA (1993) Hepatopulmonary syndrome: clinical observations and lack of therapeutic response to somatostatin analogue. Chest 104:515–521
4. Hourani JM, Bellamy PE, Tashkin DP, Batra P, Simmons MS (1991) Pulmonary dysfunction in advanced liver disease: frequent occurrence of an abnormal diffusing capacity. Am J Med 90:693–700
5. Rodriguez-Roisin R, Wagner PD (1990) Clinical relevance of ventilation-perfusion inequality determined by inert gas elimination. Eur Respir J 3:469–482
6. Rodriguez-Roisin R, Roca J, Agustí AGN, Mastai R, Wagner PD, Bosch J (1987) Gas exchange and pulmonary vascular reactivity in patients with liver cirrhosis. Am Rev Respir Dis 135:1085–1092
7. Genovesi MG, Tierney DF, Taplin GV, Eisenberg H (1976) An intravenous radionuclide method to evaluate hypoxemia caused by abnormal alveolar vessels. Am Rev Respir Dis 114:59–65
8. Agustí AGN, Roca J, Rodriguez-Roisin R, Mastai R, Wagner PD, Bosch J (1989) Pulmonary hemodynamics and gas exchange during exercise in liver cirrhosis. Am Rev Respir Dis 139:485–491

9. Wagner PD, Rodriguez-Roisin R (1991) Clinical advances in pulmonary gas exchange. Am Rev Respir Dis 143:883–888
10. Andrivet P, Cadranel J, Housset B, Herigault R, Harf A, Adnot S (1993) Mechanisms for impaired arterial oxygenation in patients with liver cirrhosis and severe respiratory insufficiency: effects of indomethacin. Chest 103:500–507
11. Rodriguez-Roisin R, Krowka MJ (1994) Is arterial hypoxaemia due to hepatic disease an indication for liver transplantation? A new therapeutic approach. Eur Respir J 7:839–842
12. Vincent JL, Lignian H, Gillet JB, Contu E (1985) Increase in PaO_2 following intravenous administration of propranolol in acutely hypoxemic patients. Chest 88:558–562
13. Agustí AGN, Roca J, Bosch J, García-Pagan JC, Wagner PD, Rodriguez-Roisin R (1990) Effects of propranolol on arterial oxygenation and O_2 transport to tissues in patients with cirrhosis. Am Rev Respir Dis 142:306–310
14. Schumacker PT, Cain SM (1987) The concept of a critical oxygen delivery. Intensive Care Med 13:223–229
15. Bihari D, Gimson AES, Lindrige J, Williams R (1985) Lactic acidosis in fulminant hepatic failure. J Hepatol 1:405–416
16. Shibutani K, Komatsu T, Kubal K, Sanchala V, Kumar V, Bizzari DV (1983) Critical level of oxygen delivery in anesthetized man. Crit Care Med 11:640–643

Regional Monitoring of Gut Mucosal Viability

G. Gutierrez and R. A. Cabezas

Introduction

It is increasingly evident that systemic measures of oxygen transport (TO_2) and consumption (VO_2) fail to provide the information necessary to reverse the disastrous consequences of multiple organ dysfunction. Perhaps our disillusion with these systemic parameters is the result of overly optimistic expectations that physiological principles of mass transport could be sufficiently sensitive to characterize the complex process of tissue oxygenation [1]. Systemic TO_2, defined as the product of cardiac output and arterial O_2 content, portrays the capacity of the cardiopulmonary system to oxygenate and transport blood to the major vessels. Systemic VO_2 is a measure of the rate of oxygen uptake by perfused tissues, the bulk of which is used by the mitochondria to produce energy via oxidative phosphorylation. These systemic parameters, however, offer scant information on the aerobic energy balance of cells.

Another concept that has gained acceptance during the past few years is that tissue perfusion during critical illness is heterogeneous. That is, not all tissues receive their fair share of oxygen and nutrients under conditions of stress. Under these circumstances, perfusion is maintained, or even increased, in organs normally associated with locomotion, i.e., the skeletal muscles, heart, brain, and adrenal glands. Perhaps this reflects a limited repertoire of responses to physical danger embodied in the flight or fight reaction. Organs that fare less well in critical illness, including systemic hypoxia and sepsis, are the gastrointestinal mucosa and the kidneys.

Unfortunately, our ability to assess changes in regional perfusion and oxygenation is severely hampered by the lack of monitoring techniques to measure local changes in tissue oxygenation and bioenergetics. Among the techniques capable of measuring metabolic alterations during cellular hypoxia are magnetic resonance spectroscopy [2], near-infrared spectroscopy [3], and gastric or intestinal tonometry [4]. Of these, only gastric tonometry is clinically available at present. This technique monitors increases in gut intramucosal PCO_2 as a marker of local anaerobiasis. The assumption that mucosal bicarbonate is approximately equal to arterial bicarbonate allows the computation of mucosal pH (pHi). This parameter has been proposed as an index of cellular energy balance of the gut mucosa, a tissue that is exquisitely sensitive to decreases in oxygenation [5, 6].

Alterations in Regional Blood Flow During Hypoxia

The fraction of the cardiac output directed to skeletal muscle increases with the degree of hypoxemia, as local microvascular vasodilation overcomes neurogenic and humoral vasoconstriction. Vogel et al. [7] exposed rabbits to an atmosphere of 8% oxygen in nitrogen and measured regional blood flow with the rubidium fractional uptake technique. They found significant increases in cardiac output and arterial blood pressure. Furthermore, the fraction of cardiac output directed to the myocardium and skeletal muscle increased, whereas that for the brain remained unchanged. Decreases in blood flow occurred in the kidneys, liver, and intestines. Doherty and Liang [8] gave phentolamine (α blockade), or a combination of phentolamine with propanolol ($\alpha + \beta$ blockade) to awake dogs exposed to decreases in FIO_2 to 0.08 and 0.05. They found a graded increase in cardiac output with the two levels of hypoxemia. Flow to the heart, brain, adrenals, and skeletal muscle increased, whereas renal and splanchnic flows were unaltered by hypoxemia. They concluded that local metabolic factors are primarily responsible for hypoxic vasodilation in the awake dog.

Nesarajah et al. [9] used radioactive microspheres in conscious, standing adult ewes breathing an FIO_2 of 0.08 for 20 min. They noted increases in cardiac output, arterial blood pressure, and increases in blood flow to most organs, except to the spleen and the intestine. The fraction of cardiac output increased for the heart, brain, diaphragm, and skeletal muscle and decreased for the intestines and splanchnic organs. The preferential distribution of blood flow during hypoxemia to the myocardium, brain, adrenals, diaphragm, and skeletal muscle as well as a decline in blood flow to the kidneys and most of the gut has been confirmed in awake rabbits [10], sheep [11], and newborn lambs [12]. Therefore, it makes sense to monitor alterations in tissue oxygenation, or, even better, in cellular energy balance, in the gut and the kidney as early warnings of systemic hypoxia. This objective appears to be feasible with gastric or intestinal tonometry.

Definition of Tonometry

The technique of tonometry refers to the equalization of partial pressures between a fluid, such as blood or saline, and a gas of known composition. The technique relies on bringing the gas and the fluid in close contact, usually by having the gas flow over a thin fluid film. After a given equilibration time the partial pressures become similar in both media. Gastric or intestinal tonometry is based on the principle of luminal fluid PCO_2 coming into equilibrium with the PCO_2 of the mucosa layer.

The earliest application of gastric tonometry can be found in the work of Boda and Murányi [13], who used tonometry in 1959 to measure mucosal PCO_2 in intubated, mechanically ventilated children with polio and used it as a surrogate for arterial PCO_2. They fashioned a tonometer out of a thin-walled rubber balloon attached to a polyethylene tube and introduced it through the nose into the stomach. The balloon was filled with air and required 2 h for equilibration. Further, they proposed the use of tonometric measures of PCO_2 combined with infrared measurement of CO_2 expired air to monitor ventilated patients. Based on their clinical experience with this device they concluded that "the CO_2 tension of arterial blood could be estimated with reasonable accuracy." Boda and Murányi also recognized that an increase in gastric PCO_2 is a sign of regional tissue hypoperfusion, since they noted in their paper that "gastric PCO_2 may be *misleadingly high* in severe shock accompanied by an excessive slowing of the circulation."

Unfortunately, the concept of tonometry did not find clinical acceptance at the time, and the work of Boda and Murányi was forgotten. Tonometry went through a process of rediscovery in the decades of the 1960s and 1970s. Bergofsky [14] used tonometry to measure the PO_2 and PCO_2 of the gallbladder and urinary bladder mucosa by injecting saline into the lumen of these organs. The PCO_2 in the saline increased, and the PO_2 decreased until they approached values near those measured in blood. The first use of tonometry in the intestine was reported by Dawson et al. [15], who also instilled saline into intestinal pouches of experimental animals and found that arterial tonometer PCO_2 approximated arterial PCO_2 after a suitable equilibration period. Kivisaari and Niinikoski [16] used a flexible saline filled tube made of a silastic, a material permeable to O_2 and CO_2, to measure the intradermal PO_2 and PCO_2 and used as indicators of wound healing.

The Concept of Intramucosal pH

In 1982 Fiddian-Green et al. [17] advanced the notion that the pH of the gut mucosa could be calculated using tonometry. They proposed that measures of tonometrically measured PCO_2 approximates intramucosal PCO_2. This assumes that CO_2 diffuses freely in tissue, so that PCO_2 in the luminal fluid is in equilibrium with the mucosal PCO_2. Another assumption made was that the bicarbonate concentration ($[HCO^-]$) of the intestinal mucosa is similar to that of arterial blood. The application of the Henderson-Hasselbalch equation to these variables yields the value of pHi:

$$pHi = 6.1 + \log \frac{HCO_3^-}{\alpha PCO_2}$$

where α represents the solubility of CO_2 in plasma ($\alpha = 0.03$). Currently, commercially available tonometers consist of a silicone balloon placed at the tip of a flexible catheter, as a nasogastric tube for use in the stomach or as a smaller diameter catheter for use in the sigmoid colon.

Physiological and Practical Aspects of Tonometry

Objections to the use of tonometry as a clinical monitor of regional tissue oxygenation revolve around the validity of the signal as a physiological index of hypoxia and the technical aspects associated with the procedure.

Physiological Aspects of Tonometry

Increases in tissue concentration of carbon dioxide occur both during normoxia and hypoxia. Cells require a continuous supply of energy to perform their assigned roles and to maintain adequate ionic equilibrium with respect to the surrounding fluid. Cellular energy is stored in the high energy phosphate bonds of adenosine triphosphate (ATP). This energy is readily released at the sites of utilization during the hydrolysis of ATP, resulting in,

$$ATP \rightarrow ADP + Pi + H^{+} + \textit{Energy}$$

Under conditions of oxygen abundance, or normoxia, the intracellular concentrations of ADP, Pi, and H^{+} are very small, as these metabolites are used in the mitochondria to reconstitute ATP in the process of oxidative phosphorylation. The aerobic production of ATP results in the consumption of O_2 and in the production of CO_2 by the mitochondria at a rate determined by the respiratory quotient ($RQ = CO_2$ production/O_2 consumption).

During hypoxia the hydrogen ions resulting from the hydrolysis of ATP are not recycled in the mitochondria but are buffered by bicarbonate. This results in an increase in tissue CO_2 beyond that ascribed to aerobic metabolism, and the cellular production of CO_2 is greater than the consumption of O_2.

$$H^{+} + HCO_3^{-} \leftrightarrow H_2O + CO_2$$

The premise that PCO_2 in the lumen is in equilibrium with mucosal PCO_2 appears to have a firm physiological foundation. As shown by Cunningham et al. [18] using mass spectrometry, there are no significant differences between luminal and extraluminal PCO_2 in rats with or without intestinal ischemia. On the other hand, the assumption of equality between mucosal and arterial $[HCO_3^{-}]$ has been questioned. Antonsson et al. [19] compared measures of pig ileal pHi obtained with a silastic balloon tonometer to those measured directly with glass pH microelectrodes. The two methods yielded similar pHi values during a 3-h control period and following the injection of *Escherichia coli* endotoxin. However, during total or partial mesenteric occlusion, tonometrically calculated pHi underestimated electrode measured pHi. The disparity in pHi measurements during total and partial ischemia in these experiments may be related to the long time constant associated with tonometry, resulting in an underestimation of pHi during changing conditions. Also, arterial [HCO_3^-] probably is greater than tissue [HCO_3^-] during mesenteric occlusion. Relative decreases in tissue [HCO_3^-] most likely take place by buffering tissue H^+.

Practical Aspects of Tonometry

In its present form tonometry is a cumbersome technique. It is labor intensive and the accuracy of the information depends to a great extent on the care with which samples are taken. The measurement of gastric or intestinal pHi entails the following sequence of events:

1. Introduce the correct amount of saline into the tonometer balloon.
2. Wait for an equilibrium period of 30–90 min.
3. Draw and discard the saline occupying the tonometer's dead space.
4. Obtain a sample of the balloon's saline using anaerobic technique.
5. Take the sample to a blood gas analyzer and measure the saline's PCO_2.
6. Determine the steady-state PCO_2 based on the equilibration time and a monogram supplied by the manufacturer.
7. Compute pHi using the Henderson-Hasselbalch equation in which the bicarbonate term is that measured an arterial blood sample.

Obviously, there are ample opportunities to introduce errors at any point of this process. Perhaps the most important sources of errors are:

1. Inaccurate timing of partially equilibrated samples. This problem can be avoided by waiting at least 90 min between samples.
2. Introduction of air in the saline sample. This falsely decreases PCO_2 and raises pHi.
3. The use of blood gas analyzers that introduce a large bias in the measurement of saline PCO_2. This problem can be corrected by using a phosphate bicarbonate buffer [20] and using proper blood gas analyzers.
4. Failure to use H_2 receptor blockers when using gastric tonometers. The latter moderate increases in luminal H^+ in the stomach, whose buffering by exogenous or endogenous bicarbonate, may elevate PCO_2, thus falsely lowering pHi [21].

Physiological Aspects of Tonometry

It has been proposed that, as a consequence of impaired tissue CO_2 removal by the circulation, tissue PCO_2 can increase in the absence of tissue dysoxia. Therefore the argument follows that increases in mucosal PCO_2 merely reflect impaired CO_2 clearance, instead of the buffering of anaerobically produced H^+. This notion has its basis on the work of Weil and coworkers [22, 23] and Adrogue et al. [24], who noted increases in venous PCO_2 and decreases in venous pH in patients undergoing resuscitation from cardiac arrest, and suggested that these changes are the result of respiratory rather than metabolic alterations. This may well be the case during a cardiac arrest, as O_2 stores are consumed, and CO_2 is produced but not removed from the tissues. However, complete cessation of blood flow, a condition that does not require a tonometer for diagnosis, is relatively rare in critically ill patients. As shown by Roukonen et al. [25] in patients following cardiac surgery, splanchnic flows and hepatic vein lactate were similar in a group with low cardiac index ($CI < 1.75\ l\ min^{-1}\ m^{-2}$) when compared to a control group ($CI \approx 2.5\ l\ min^{-1}\ m^{-2}$). In a separate study these investigators also showed that gut dysoxia in patients with sepsis may be related to increases in gastrointestinal metabolic rate rather than to decreases in splanchnic blood flow [26].

The issue of whether decreasing intestinal pHi reflects metabolic acidosis or simply stagnation of CO_2 produced by oxidative phosphorylation as flow decreases was addressed experimentally by Schlichtig and Bowles [27]. They subjected dogs to progressive decreases in intestinal blood flow during cardiac tamponade and measured pHi, intestinal O_2 delivery (DO_2), VO_2, and the arterio-venous differences (a-v) in blood pH and PCO_2. They found that the critical DO_2, defined as the minimum DO_2 capable of sustaining aerobic metabolism, coincided with decreases in pHi. Further, the increment in $pH_{a\text{-}v}$ per increment in $\log(PCO_2)_{a\text{-}v}$ was significantly greater at DO_2 values below the critical DO_2, implying that decreases in intestinal pHi during progressive flow reduction are associated with the onset of tissue metabolic acidosis. In a subsequent study using a similar animal model [28], these investigators found that the hemoglobin saturation of portal blood calculated with the Dill nomogram from PCO_2 agreed with the measured level of oxygenated hemoglobin (HbO_2). This is to be expected, given the linear relationship between PCO_2, CO_2 content, and the arteriovenous O_2 content. However, below the critical O_2 transport point the relationship between HbO_2 and PCO_2 was no longer linear as mucosal PCO_2predicted values of HbO_2 lower than those measured in portal blood. This indicated anaerobic dissolved CO_2 production by the intestinal mucosa. Similar findings were reported by Zhang and Vincent [29] in dogs subjected to progressive pericardial tamponade. They noted considerable increases in systemic $pH_{a\text{-}v}$ and $(PCO_2)_{a\text{-}v}$ and in arterial blood lactate concentration at DO_2 values below critical DO_2. They also concluded that increases in $pH_{a\text{-}v}$ and $(PCO_2)_{a\text{-}v}$ represent reliable parameters of tissue hypoxia.

Experimental and Clinical Experience with Tonometry

Intestinal Ischemia

Increases in mucosal PCO_2 or conversely, decreases in pHi are associated with the development of intestinal mucosa ischemia. Rabbits subjected to mesenteric artery occlusion show rapid and sustained increases in intraluminal PCO_2 that correlate strongly with the histologic grade of injury [30]. Grum et al. [31] decreased mesenteric blood flow in dogs in a step-wise fashion and noted that pHi remained constant until a critical O_2 delivery was reached, when it began to decline in concert with regional VO_2. The clinical utility of pHi to detect intestinal mucosal ischemia was tested by Fiddian-Green et al. [32] in 25 high-risk patients undergoing abdominal aortic surgery. Six patients with sigmoid colon pHi below 6.86 developed evidence of ischemic colitis, whereas the other 19 patients, who had pHi values above 6.86 did not. In a prospective study of 34 patients undergoing elective and emergent aortic reconstruction Schiedler et al. [33] also found a significant correlation between decreases in pHi and the development of ischemic colitis. Decreases in pHi also have been associated with chronic gastric [34] and mesenteric ischemia [35]. More recently Björck and Hedberg [36] found that pHi below 6.86 predicted endoscopically ischemic colitis after abdominal aortic surgery, and that pHi below 7.10 for longer than 2 h served as an early warning of postoperative complications (sensitivity 100%, specificity 92%).

Intestinal Translocation

Fiddian-Green and Gantz [37] noted that patients undergoing abdominal aortic operations, and who subsequently developed infection from intestinal organisms, had lower sigmoid colon pHi. They related the probability of infection to the duration of the colonic ischemia, suggesting that there may be a relationship between the magnitude and duration of mucosal acidosis and the translocation of organisms from the gut. Andersen et al. [38] also noted decreases in gastric pHi in patients undergoing cardiopulmonary bypass, but found no correlation between gastric pHi and the levels of endotoxin, endotoxin antibody, or tumor necrosis factor-α in blood.

pHi as Predictor of ICU Morbidity and Mortality

A low gastric mucosal pHi on admission to the ICU appears to be predictive of mortality. This was confirmed by Gys et al. [39] who measured gastric pHi by gastric fluid aspiration in of 59 postsurgical subjects upon admission to the ICU. Patients with admission pHi below 7.32 had a 37% short-term mortality rate (defined as death within 72 h of ICU admission) compared to 0% in those with pHi of 7.32 or higher. They also found that 90% of patients with sepsis had pHi levels lower than 7.32. They found no correlation between gastric pHi measured on ICU admission and hospital mortality.

Doglio et al. [40] measured gastric pHi with a tonometer in a heterogeneous group of 80 patients at the time of admission to the ICU and again 12 h later. A pHi of 7.35 was used as the lower range of normality. As shown in Fig. 1, the group admitted with low pHi had greater ICU mortality (65% vs.

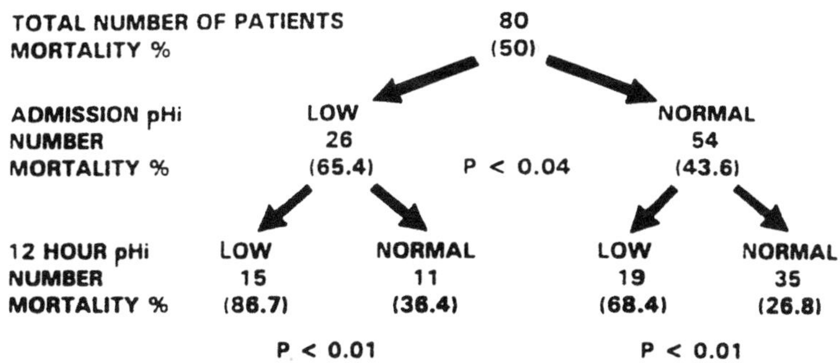

Fig. 1. Relationship of mortality rate to measurements of gastric mucosal pH (*pHi*) made at the time of admission to the intensive care unit and again 12 h later. Patients with a persistently low pHi, less than 7.35, had a mortality rate of 87.6%. Those with pHi greater than or equal to 7.35 had a mortality rate of 26.8%. (From [40])

44%; $p<0.04$). Further, those patients with persistently low pHi 12 h after ICU admission had the highest mortality rate (87%), compared to a mortality rate of only 27% for the group with normal pHi during the 12 h following ICU admission. Also, the prevalence of sepsis was greater in the low-pHi group (59% vs. 26%; $p<0.01$). This study was repeated by Maynard et al. [41], who found a remarkably concordance in the data from the two studies. They compared the predictive ability of pHi to that of lactate concentration, DO_2, and VO_2 among others, and found that gastric pHi had a very high sensitivity (88%) and the highest likelihood ratio of the variables tested. Further analysis of the data using a step-down logistic regression analysis showed that pHi at 24 h was the only independent predictor of mortality. Of interest, this study was conducted without the use of H_2 receptor blocking agents.

In a prospective study comparing measures of gastric pHi to standard measures of TO_2 and VO_2 in a group of 22 patients [42], only a low gastric pHi and low mixed venous pH could separate survivors from those who died during the study. Marik [43] obtained similar results in a study of 30 ICU patients, in whom gastric pHi and mixed venous pH were lower in patients who developed multiple systems organ failure and in those who died. Roumen et al. [44] measured pHi in patients during the 48 h following multiple blunt trauma. They found that seven patients with pHi of 7.32 or higher had uncomplicated recoveries, whereas three out of eight patients with pHi below 7.32 developed major complications, and two of them subsequently died. They found no correlation between pHi and base deficits, APACHE II scores, lactic acidosis, or Injury Severity Score values.

Gastric pHi as an Index of Resuscitation

Changes in pHi have been used to guide therapy in critically ill patients. In a prospective trial, Gutierrez and colleagues [45] studied 260 patients admitted

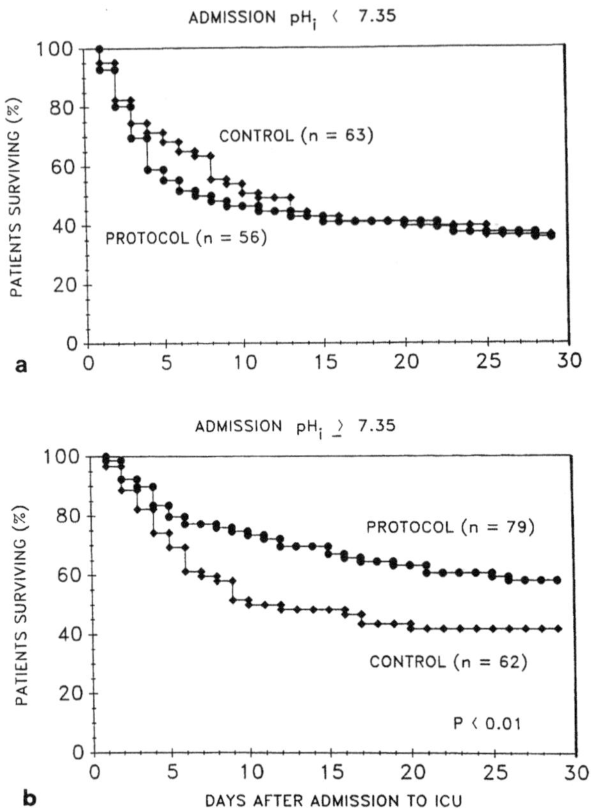

Fig. 2a, b. Kaplan-Meier curves for patients treated according to a pHi-guided protocol aimed at increasing oxygen delivery. Patients admitted to the intensive care unit with gastric mucosal pH (pHi) less than 7.35 (**a**) did not benefit from the treatment protocol, whereas those with admission pHi greater than 7.35 (**b**) showed a significant reduction in mortality when treated according to the pHi-guided protocol. (From [45])

to the ICU of eight participating hospitals with APACHE II scores of 15–25. After insertion of a gastric tonometer each patient was randomized to a control or protocol group within an admission category (normal ≥ .35; low < 7.35). The control groups were treated according to standard ICU practices, whereas the protocol group received, in addition, treatment to increase systemic DO_2 whenever pHi decreased below 7.35 or by more than 0.10 units from the previous measurements. The intervention protocol consisted of standard treatment of hypotension, hypoxia, anemia, fever, etc., followed by the infusion of fluids and the inotrope dobutamine. The protocol was used in all patients admitted with low pHi and, because pHi fell at some time during their ICU stay, in 85% of those admitted with normal pHi. They found that for patients admitted with pHi below 7.35 survival was similar in the protocol and control groups (37% vs. 36%; Fig. 2), whereas for those admitted with normal pHi, survival was significantly greater in the protocol treated group (58% vs. 42%; $p<0.01$). The authors concluded that pHi-guided resuscitation may improve outcome in the subpopulation of patients admitted to the ICU with normal pHi by preventing splanchnic ischemia and the development of a systemic oxygen deficit.

This study has been criticized for the apparent lack of a well-defined control population, since patients in that group were treated according to local standards of care rather than using a common treatment protocol. Since patients in the protocol groups received a greater number of interventions than those in the control groups, it may be that the difference between the groups was the result of a Hawthorne type effect (where improvements in outcome are the result of a greater level of medical care and not of a specific therapeutic effort). This possibility was acknowledged by the investigators, who measured scores on the Therapeutic Intervention Scoring Sytem during the time of the study. They found that patients in the low pHi groups required a greater level of effort, especially during the first 48 h following admission to the ICU, but there were no significant differences between the control and protocol groups for each category of pHi. Moreover, the nearly identical mortality rate in the low-pHi group, irrespective of randomization to the control or to the protocol group, also seems to rule out a treatment bias. The fact that patients in the protocol group received a significantly greater number of interventions directed at increasing systemic O_2 transport was precisely the point of the study, which called for these interventions whenever pHi was less than 7.35. As it stands, this study represents the only prospective trial in the literature testing the efficacy of a monitoring device to direct therapy. Obviously, other prospective studies on the use of pHi-guided therapy, directed at specific patient populations, and perhaps employing different treatment modalities should be undertaken.

Other Clinical Applications of Gastric Tonometry

Measures of gastric pHi, obtained by sampling gastric juice directly, have been shown by Mohsenifar et al. [46] to be predictive of weaning success in mechanically ventilated patients. They found that those patients that could not be weaned from assisted mechanical ventilation experienced significant decreases in pHi during weaning, 7.09 compared to 7.36 during mechanical ventilation, whereas patients who were successfully weaned showed no change in pHi. They concluded from the study that pHi may be an early sign of weaning failure as blood flow is diverted to the respiratory muscles during weaning, away from the splanchnic bed and other "nonvital" tissues. In another study Marik and Sibbald [47] were able to establish an inverse association between changes in gastric pHi in septic patients and the age of transfused blood. Patients receiving blood stored for more than 15 days experienced decreases in gastric mucosal pHi in a consistent basis, perhaps implying that poorly deformable transfused red blood cells cause microcirculatory occlusion in some organs.

Gastric tonometry also has been used to predict the likelihood of bleeding from stress ulcerations. Fiddian-Green et al. [48] studied 103 ICU patients and found massive bleeding from stress ulcerations in seven patients. The mean pHi in these patients was 7.02 moreover, bleeding was not observed in patients with pHi higher than 7.24, although the pH of the gastric juice was similar for all groups at approximately 5.0. Similar data were reported by Diebel et al. [49] in patients with chronic renal failure on maintenance dialysis. They also reported in this study normal pHi values in a group of seven healthy volunteers as 7.37 ± 0.04 and a normal arterial-mucosal pH difference of 0.02 ± 0.01.

Decreases in gastric pHi have been used as an indicator of allograft viability in orthotopic liver transplants. Frenette et al. [50] found that patients with pHi of 7.30 or higher after reperfusion during the neohepatic phase of orthotopic liver transplantation, and those in whom pHi returns to normal within 3 h after reperfusion fared well during this procedure. However, a patient who had pHi lower than 7.30 after reperfusion, and whose pHi remained low, had to undergo retransplantation the following day. In another study of patients undergoing orthotopic liver transplantation Downing et al. [51] found increased incidence of sepsis and graft dysfunction in a groups of patients who had pHi values below 7.32 during the procedure. These are small series and further studies are needed to confirm the utility of pHi during liver transplantations.

Conclusions

Monitoring increases in gut mucosal PCO_2 appears to be very useful in the assessment of regional tissue oxygenation in critically ill patients. The physiological signal provided by tonometry appears to correlate with local alterations in cellular energy balance and to provide an early warning when levels of systemic oxygen transport may not be sufficient to satisfy the needs of all organs.

Acknowledgements. This research was supported in part by an American Lung Association Career Investigator Award.

References

1. Gutierrez G, Vincent JL (1994) Meeting report. Round table conference on tissue oxygenation in ARDS. Clin Intensive Care 4:300–301
2. Gutierrez G (1991) Cellular energy metabolism during hypoxia. Crit Care Med 19:619
3. Piantadosi CA, Hemstreet TM, Jöbsis-Vandervliet FF (1986) Near-infrared spectrophotometric monitoring of oxygen distribution to intact brain and skeletla muscle tissues. Crit Care Med 14:698–706
4. Clark CH, Gutierrez G. Gastric intramucosal pH (1992) A noninvasive method for the indirect measurement of tissue oxygenation. Am J Crit Care 1:53–60

5. Fiddian-Green RG (1991) Editorial: should measurements of tissue pH and PO_2 be included in the routine monitoring of intensive care unit patients? Crit Care Med 19:141–143
6. Dantzker D (1993) Editorial: the gastrointestinal tract: the canary of the body? JAMA 270:1247–1248
7. Vogel JA, Pulver RI, Burton TM (1969) Regional blood flow distribution during simulated high-altitude exposure. Fed Proc 28:1155–1159
8. Doherty JU, Liang CS (1984) Arterial hypoxemia in awake dogs. J Lab Clin Med 104:665–677
9. Nesarajah M, Matalon S, Krasney JA, Farhi LE (1983) Cardiac output and regional oxygen transport in the acutely hypoxic conscious sheep. Respir Physiol 53:161–172
10. Wyler F (1975) Effects of hypoxia on distribution of cardiac output and organ blood flow in the rabbit. Cardiology 60:163–172
11. Davidson D, Stalcup SA (1984) Systemic circulatory adjustments to acute hypoxia and reoxygenation in unanesthetized sheep. J Clin Invest 73:317–328
12. Fisher DJ (1984) Cardiac output and regional blood flows during hypoxaemia in unanesthetized newborn lambs. J Dev Physiol 6:485–494
13. Boda D, Murányi L (1959) Gastrotonometry. An aid to the control of ventilation during artificial respiration. Lancet 273:181–182
14. Bergofsky EH 81964) Determination of tissue O_2 tensions by hollow visceral tonometers: effect of breathing enriched O_2 mixtures. J Crit Invest 43:193–200
15. Dawson AM (1965) Small bowel tonometry: assessment of small gut mucosal oxygen tension in dog and man. Nature 206:943–944
16. Kivisaari J, Niinikoski J (1973) Use of silastic tube and capillary sampling technic in the measurement of tissue PO_2 and PCO_2. Am J Surg 125:623–627
17. Fiddian-Green RG, Pittenger G, Whitehouse WM (1982) Back-diffusion of CO_2 and its influence on the intramural pH in gastric mucosa. J Surg Res 33:39–48
18. Cunningham JA, Cousar CD, Jaffin JH, Harmon JW (1987) Extraluminal and intraluminal PCO_2 levels in the ischemic intestines of rats. Curr Surg 44:229–232
19. Antonsson JB, Boyle CC, Kruithoff KL et al (1990) Validation of tonometric measurement of gut intramural pH during endotoxemia and mesenteric occlusion in pigs. Am J Physiol 259:G519–G523
20. Takala J, Parviainen I, Siloaho M, Ruokonen E, Hämäläinen E (1994) Saline PCO_2 is an important source of error in the assessment of gastric mucosal pH. Crit Care Med 22:1877–1879
21. Kolkman JJ, Groeneveld ABJ, Meuwissen SGM (1994) Effect of ranitidine and bicarbonate enhanced intragastric PCO_2: a tonometric study. Gut 35:737–741
22. Weil MH, Rackow EC, Trevino RP, Grundler W, Falk JL, Griffel MI (1986) Difference in acid-base state between venous and arterial blood during cardiopulmonary resuscitation. N Engl J Med 315:153–156
23. Grundler W, Weil MH, Rackow RC (1986) Arteriovenous carbon dioxide and pH gradients during cardiac arrest. Circulation 74:1071–1074
24. Adrogue JH, Rashad MN, Gorin AB, Yacoub J, Madias NE (1989) Assessing acid-base status in circulatory failure: difference between arterial and central venous blood. N Engl J Med 320:1312–1316
25. Ruokonen E, Takala J, Kari A (1992) Regional blood flow and oxygen transport in patients with the low cardiac output syndrome after cardiac surgery. Crit Care Med 21:1304–1311
26. Ruokonen E, Takala J, Kari A, Saxen H, Mertsola J, Hansen E. (1992) Regional blood flow and oxygen transport in septic shock. Crit Care Med 21:1296–1303
27. Bowles SA, Schlichtig R, Kramer DJ, Klions HA (1992) Arteriovenous pH and partial pressure of carbon dioxide detect critical oxygen delivery during progressive hemorrhage in dogs. J Crit Care 7:95–105
28. Schlichtig R, Bowles SA (1994) Distinguishing between aerobic and anaerobic appearance of dissolved CO_2 in intestine during low flow. J Appl Physiol 76:2443–2451

29. Zhang H, Vincent JL (1993) Arteriovenous differences in PCO_2 and pH are good indicators of critical hypoperfusion. Am Rev Respir Dis 148:867–871
30. Fiddian-Green RG, Stanley JC, Nostrant T et al (1989) Chronic gastric ischemia: a cause of abdominal pain or bleeding identified from the presence of gastric mucosal acidosis. J Cardiovasc Surg 852–859
31. Grum CM, Fiddian-Green RG, Pittenger GL et al (1984) Adequacy of tissue oxygenation in intact dog intestine. J Appl Physiol 56:1065–1069
32. Fiddian-Green RG, Amelin PM, Herrmann JB et al (1986) Prediction of the development of sigmoid ischemia on the day of aortic operations. Arch Surg 121:654–660
33. Schiedler MG, Cutler BS, Fiddian-Green RG (1987) Sigmoid intramural pH for prediction of ischemic colitis during aortic surgery. Arch Surg 122:881–886
34. Fiddian-Green RG, Stanley JC, Nostrant T et al (1989) Chronic gastric ischemia: a cause of abdominal pain or bleeding identified from the presence of gastric mucosal acidosis. J Cardiovasc Surg 852–859
35. Boley SJ, Brandt LJ, Veith FJ et al (1991) A new provocative test for chronic mesenteric ischemia. Am J Gastroenterol 86:888–891
36. Björk M, Hedberg B (1994) Early detection of major complications after abdominal aortic surgery: predictive value of sigmoid colon and gastric intramucosal pH monitoring. Br J Surg 81:25–30
37. Fiddian-Green RG, Gantz NM (1987) Transient episodes of sigmoid ischemia and their relation to infection from intestinal organisms after abdominal aortic operations. Crit Care Med 15:835–839
38. Andersen LW, Landow L, Baek L et al (1993) Association between gastric intramucosal pH and splanchnic endotoxin, antibody to endotoxin, and tumor necrosis factor-α concentrations in patients undergoing cardiopulmonary bypass. Crit Care Med 21:210–217
39. Gys T, Hubens A, Neels H et al (1988) Prognostic value of gastric intramural pH in surgical intensive care patients. Crit Care Med 16:1222–1224
40. Doglio GR, Pusajo JF, Egurrola MA et al (1991) Gastric mucosal pH as a prognostic index of mortality in critically ill patients. Crit Care Med 19:1037–1040
41. Maynard N, Bihari D, Beale R et al (1993) Assessment of splanchnic oxygenation by gastric tonometry in patients with acute circulatory failure. JAMA 270:1203–1210
42. Gutierrez G, Bismar H, Dantzker DR et al. (1992) Comparison of gastric intramucosal pH with measures of oxygen transport and consumption in critically ill patients. Crit Care Med 20:451–457
43. Marik PE (1993) Gastric intramucosal pH: a better predictor of multiorgan dysfunction syndrome and death than oxygen-derived variables in patients with sepsis. Chest 104:225–229
44. Roumen RM, Vreugde JPC, Goris JA (1994) Gastric tonometry in multiple trauma patients. J Trauma 36:313–316
45. Gutierrez G, Palizas F, Doglio G et al (1992) Gastric intramucosal pH as a therapeutic index of tissue oxygenation in critically ill patients. Lancet 339:195–199
46. Mohsenifar Z, Hay A, Hay J et al (1993) Gastric intramural pH as a predictor of success or failure in weaning patients from mechanical ventilation. Ann Intern Med 119:794–798
47. Marik PE, Sibbald WJ (1993) Effect of stored-blood transfusion on oxygen delivery in patients with sepsis. JAMA 269:3024–3029
48. Fiddian-Green RG, McGough E, Pittenger G et al (1983) Predictive value of intramural pH and other risk factors for massive bleeding from stress ulceration. Gastroenterology 85:613–620
49. Diebel L, Kozol R, Wilson RF et al (1993) Gastric intramucosal acidosis in patients with chronic kidney failure. Surgery 113:520–526
50. Frenette L, Doblar DD, Singer D, Cox J, Ronderos J, Poplawski S, Ranjan D (1994) Gastric intramural pH as indicator of early allograft viability in orthotopic liver transplantation. Transplantation 3:292–297
51. Downing A, Cottam S, Beard C, Potter D (1993) Gastric mucosal pH predicts major morbidity following orthotopic liver transplantation. Transpant Proc 2:1804

Gastric Intramucosal pH Determination: Limitations of the Technique

X. Calvet, F. Baigorri, and A. Artigas

Introduction

Gastric intramucosal pH (pHi) measurement by means of tonometry is a relatively noninvasive monitoring technique that measures intestinal perfusion indirectly. Briefly, a nasogastric tube with a silicone balloon permeable to CO_2 at its tip is introduced into the gastric lumen. The balloon is filled with saline, CO_2 is allowed to equilibrate with mucosal CO_2, and saline is then aspirated and the PCO_2 measured. pHi is calculated according to Henderson-Hasselbalch formula, using intramucosal PCO_2 and arterial bicarbonate.

Low pHi values are correlated with inadequate O_2 supply to the bowel, leading to regional acidosis. Intramucosal acidosis has been related to both poor prognosis and the appearance of multiorgan failure in septic patients, even in the absence of systemic acidosis or hypotension. Several studies have shown that pHi measurement by tonometry is a useful prognostic tool in critically ill patients [1–8]. Moreover, it has been suggested that the correction of intramucosal acidosis increases survival in critically ill patients [9].

Despite the progressive use of tonometry in the intensive care units some methodological aspects still remain controversial, and factors influencing in vivo reliability and variability of pHi measurements have not been completely evaluated.

The measurement of pHi by tonometry rests on two key assumptions [10, 11]: firstly, that PCO_2 in saline in the tonometer balloon is similar to tissue PCO_2 and, secondly, that tissue and arterial bicarbonate are similar. With reference to the first assumption, two groups of factors can influence the variability of PCO_2 measurements: those related to the patient and treatment, and those related to the measurement technique. The patient-related factors are the effect of gastric secretion, the use of enteral nutrition or enteral drugs, and changes in arterial PCO_2. The most important technical problems in measuring PCO_2 are the high variability of PCO_2 measurements in saline and the long equilibration period required to reach steady state which precludes rapid asessment during therapeutic interventions. With reference to the second assumption, that is, that arterial bicarbonate equals tissue bicarbonate, the relationship between arterial and tissue bicarbonate has been defined in experimental studies [12]. At the end of this chapter we review two general questions about pHi measurements by tonometry: what is normal pHi, and how reliable are pHi measurements?

Variability of PCO_2 Measurements

Patient-Related Variability

Effect of Acid Gastric Secretion on pHi. Fiddian Green et al. [13] studied the effect of gastric secretion on pHi determined by tonometry. They instilled HCl into an isolated dog stomach and measured pHi by tonometry. Luminal pH decreased from 3.0 to 1.5 following HCl, but mucosal pHi did not change. In another set of experiments with the same preparation, HCl was buffered with HCO_3^- and increases in luminal PCO_2 to values greater than 1000 mmHg were found. This resulted in a decrease in pHi from 7.31 to 6.30. Also, Kolkman et al. [14] observed that gastric intraluminal PCO_2 increased after oral bicarbonate administration in healthy volunteers. However, when acid gastric secretion was inhibited using ranitidine pretreatment, oral bicarbonate did not induce changes in intraluminal PCO_2. From these experiments it can be inferred that the back-diffusion of H^+ is not a major problem in determining pHi, whereas increases in luminal PCO_2 do confound the measurement. Therefore, CO_2 generating antiacids should not be used in measuring pHi, and, furthermore, receptor blocking agents may be useful in reducing the variability in pHi measurements by minimizing the intraluminal CO_2 production from the interaction between bicarbonate and H^+.

Effect of Ranitidine on pHi. Heard et al. [15] studied the effect of ranitidine on pHi measurements. They gave ranitidine, an H_2 receptor blocking agent, to human volunteers and measured pHi with a gastric tonometer. Compared to an untreated control group the subjects given ranitidine had lower intraluminal PCO_2, greater luminal pH, and greater mucosal pHi values. Furthermore, the variance in measurements was greater in the untreated group. From these data the administration of H_2 receptor blocking agents in conjunction with gastric tonometry was recomended. However, these results have not been validated in critically ill patients, and in this respect some clinical studies have shown that pHi measurement has a strong prognostic value despite not using H_2 blocking agents routinely before measurements [4, 5].

Preliminary results [16] on 22 critically ill patients showed that ranitidine administration significantly increases intraluminal pH but does not induce changes in measured gastric intramucosal PCO_2 (Fig. 1). Furthermore, no differences in intramucosal PCO_2 or pHi were observed between the ranitidine group and a placebo-treated group. Also, Maynard et al. [17] measured pHi before and after the administration of a single dose of 100 mg ranitidine to 12 critically ill patients. No changes in pHi (7.29 vs. 7.28) were observed

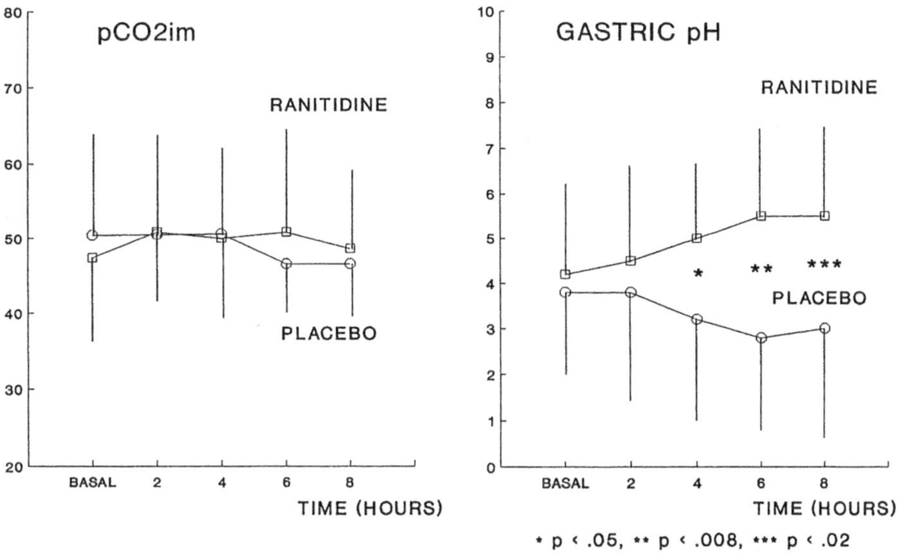

Fig. 1. Changes in intramucosal pCO2 (*pCO2im*) and in gastric intraluminal pH in the ranitidine and placebo groups

despite a marked increase in gastric luminal pH (2.1 vs. 6.5). Critically ill patients seem to respond differently than healthy volunteers to gastric acid secretion. It has been shown that critically ill patients may have an impaired acid gastric secretion [18–20]. Furthermore, Higgins et al. [20] observed that patients with low pHi had not only a basal luminal pH over 4 but also a markedly impaired pentagastrin-stimulated gastric acid secretion. As gastric acid secretion is an active metabolic process that requires an adequate supply of oxygen, these authors suggest that limited availability of oxygen to gastric mucosa can produce hypoclorhidria. In conclusion, it remains to be elucidated whether pHi may be accurately determined without ranitidine administration during gut perfusion changes in critically ill patients.

Effect of Sucralfate on pHi. It has been suggested that administration of H_2 receptor blocking agents might increase the risk of gastric bacterial colonization and nosocomial pneumonia in mechanically ventilated patients [21, 22]. In most cases, when anti-H_2 antagonists are not used, sucralfate is the preferred drug for gastrointestinal bleeding prophylaxis as it does not inhibit gastric acid secretion and does not promote bacterial colonization [23]. Despite the frequent use of sucralfate in critically ill patients there are few published results concerning the possible effect of this drug on pHi measurements. A preliminary study on 14 critically ill patients [24] found no differences between measurements of pHi and intramucosal PCO_2 performed prior to and after sucralfate administration. Also, gastric juice pH did not change after sucralfate administration (Fig. 2). Therefore, it seems that su-

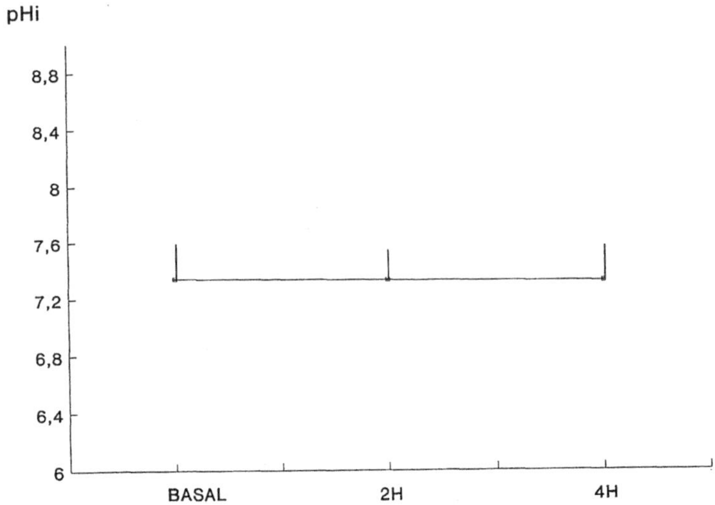

Fig. 2. Changes in pHi in patients receiving sucralfate (means). There were no significant changes from basal values

cralfate administration does not alter pHi measurements in critically ill patients.

Effect of Enteral Nutrition on pHi. There are many ways in which enteral nutrition could theoretically alter pHi measurements. Enteral nutrition might affect diffusion of CO_2 across the balloon membrane, precluding correct equilibration. Variable amounts of CO_2 might be produced from the combination of enteral nutrition and gastric acid. Increased gastric colonization might lead to intragastric CO_2 production from bacterial metabolism. Enteral feeding might induce increases in both acid and bicarbonate secretion, leading to intraluminal PCO_2 production. Finally, the metabolic requirements of digestion might increase O_2 splanchnic demand, thus aggravating gut dysoxia and lowering pHi [25–27]. Although there are no clinical data on the effect of enteral feedings on pHi, enteral nutrition has been avoided in studies evaluating pHi measurements because of these potential problems.

Effect of Changes in Arterial PCO_2 on pHi. Salzman et al. [28], in a recent report on "balloonless" tonometry induced changes in pCO_2 by variating ventilation in a group of anesthetised pigs and observed a strong correlation ($r^2=0.97, p< 0.01$) between arterial and intramucosal PCO_2. The use of the CO_2 gradient (intramucosal-arterial PCO_2) or some derived parameters, such as the pH gradient (arterial-intramucosal pH) [29] or the standarized pHi [30] rather than pHi, have been proposed as an index of splanchnic perfusion. These are thought to avoid the influence of respiratory or metabolic acid-base disturbances on pHi measurements. Standarized pHi is calculated

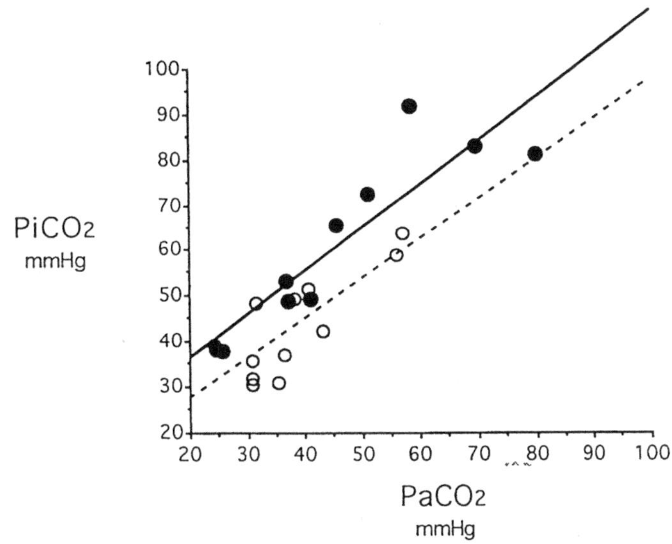

Fig. 3. Correlation between intramucosal and arterial PCO_2 in survivors (*open circles, broken line*) and nonsurvivors (*filled circles, continuous line*). Survivors: $n=12$, $r=0.79$, $p<0.02$, slope 0.87; nonsurvivors: $n=10$, $r=0.89$, $p<0.0004$, slope 0.95

using the formula: standarized pH = 7.40–(arterial pH–intramucosal pH); or: standarized pH = 7.40–pH gradient.

We have confirmed the striking correlation between intramucosal and arterial PCO_2 in a group of critically ill patients. Among these patients those with hypercapnia had significantly lower values of pHi and arterial pH. In contrast, PCO_2 gradient, pH gradient, and standarized pH did not differ significantly (Table 1). The CO_2 gradient was, however, wider in nonsurvivors than in survivors (Fig. 3), and all of these parameters – PCO_2 gradient, pHi gradient, and standarized pHi – were useful in discriminating between survivors and nonsurvivors (Table 1). These results suggest that PCO_2, pH gradient, and standarized pHi are all independent of changes in arterial PCO_2 and may be useful tools in monitoring splanchnic blood flow in critically ill patients with alterations of acid-base equilibrium. Nevertheless, the clinical value of these parameters in managing the patient and/or predicting outcome requires elucidation.

Variability of pHi Measurements Related to the Technique

Reliability of PCO_2 Measurements in Saline. It has been demonstrated that measuring PCO_2 in saline using conventional blood gas analyzers induces a considerable bias [31–33]. Thus, measurements can underestimate PCO_2 values from 15%–50% depending on the type of gas analyzer (Table 2). In fact, manufacturers of gastric tonometers recommend a specific calibration for

Table 1. Differences between patients with and without increased arterial PCO_2 and between survivors and nonsurvivors

	$paCO_2 < 45$ mmHg ($n=15$)	$paCO_2 > 45$ mmHg ($n=7$)	p	Survivors ($n=12$)	Nonsurvivors ($n=10$)	p
pHa	7.46 ± .08	7.34 ± 0.1	0.01	7.45 ± .06	7.38 ± .13	NS
$paCO_2$ (mmHg)	34 ± 6	60 ± 12	0.0001	38 ± 10	47 ± 18	NS
$CO3H^-$ (mmol/l)	24 ± 5	31 ± 6	0.005	26 ± 6	26 ± 7	NS
BE (mmol/l)	0.1 ± 5.1	5.6 ± 6.6	0.05	2.2 ± 5.6	1.4 ± 6.9	NS
$PiCO_2$ (mmHg)	42 ± 8	74 ± 12	0.0001	43 ± 11	62 ± 19	0.01
pHi	7.38 ± .12	7.24 ± .12	0.05	7.40 ± .11	7.25 ± .11	0.005
pHa–pHi	0.08 ± .08	0.1 ± .07	NS	0.05 ± .07	0.13 ± .06	0.01
pHi_{st}	7.32 ± .08	7.3 ± .07	NS	7.35 ± .07	7.27 ± .06	0.01
$PiCO_2$–$PaCO_2$ (mmHg)	8 ± 7	14 ± 12	NS	5 ± 7	15 ± 9	0.01

BE, Base excess; pHi_{st}, standardized pHi.

Table 2. Measurement bias (%) of PCO_2 in various media

Source	Gas analyzer	Saline	Gelatin	Blood	Albumin	Hartman	PBS
Riddington et al. [31]	ABL 300	−18/+9	−4/−10	−2/0			
	Nova Stat Profile 7	−68/−38	−16/−23	−3/+2			
Riddington et al. [32]	ABL 300	−18/+9	−7/+10	−2/0	−6/+10	−8/+24	
	Nova Stat Profile 7	−66/−39	−25/−17	−3/+2	+15/+45	−55/−24	
Knichwitz et al. [33]	ABL	−16/−3					−9/+3
	Ciba 288	−30/−25					−11/−8
	Nova Stat 9	−36/−58					−9/−14

saline to measure pHi. It has been suggested that variability can be reduced by using glycerine solutions [32] or phosphate-buffered solution (PBS) [33] as the CO_2 medium (Table 2). However these solutions are not widely applied at the moment. Therefore to analyze the results of a given trial it is necessary to consider the gas analyzer model employed. This is especially important in multicenter studies, where the same value of intramucosal PCO_2 or pHi may have very different meanings depending on the gas analyzer of the individual centers.

Stabilization Period and Reliability. The tonometric measurement of intramucosal PCO_2 asumes that mucosal and tonometer fluid PCO_2 are in equilibrium. Complete equilibration of balloon saline and mucosal PCO_2 requires at least 90 min. Correction factors have been developed to be applied if the measurement time is shorter than the equilibration time. Equilibration follows Fick's law and is independent of the PCO_2 but depends on factors that can influence the permeability to CO_2 such as the exchange surface, the characteristics of the balloon, and the gastric contents [27]. However it is not known to what extent these factors may influence the reliability of the measurements. Although there is no information about the effect of the equilibration period on the error of measurement, the minimum equilibration time employed has been 30 min, as variability in measurement seems to be very significant with shorter equilibration times. This is one of the major drawbacks of pHi measurements. As it is necessary to maintain the saline in the tonometer for at least 30 min, this precludes rapid monitoring of pHi changes induced by different therapeutic interventions or following accurately rapid changes in mucosal pH. Recently Saltzman et al. [28] showed that PCO_2 can be measured from the air in the stomach chamber, and that this is closely correlated with the PCO_2 in the balloon saline. As this method does not require the silicone balloon, equilibration times can probably be reduced, although this topic needs further investigation.

Variability of Bicarbonate Measurements

As noted in the "Introduction," the validity of pHi measurement by tonometry rests on two key assumptions: that PCO_2 in saline in the tonometer balloon is similar to tissue PCO_2 and that tissue and arterial bicarbonate are similar. With regard to the second assumption, a recent study by Schlichtig and Bowles [12] has demonstrated that, in an experimental dog model, arterial bicarbonate equals tissue bicarbonate until hypoxia triggers local anaerobic metabolism. At that moment bicarbonate begins to be consumed locally, and arterial bicarbonate becomes higher than tissue bicarbonate. The consumption of bicarbonate in the absence of vascular occlusion was moderate (approximately 1–4 mEq/l). Antonson et al. [10], observed that in pigs with endotoxic shock, pHi calculated by tonometry was lower than pHi directly measured by microelectrodes in all the measurements over the study period, although the differences were small and did not reach statistical sig-

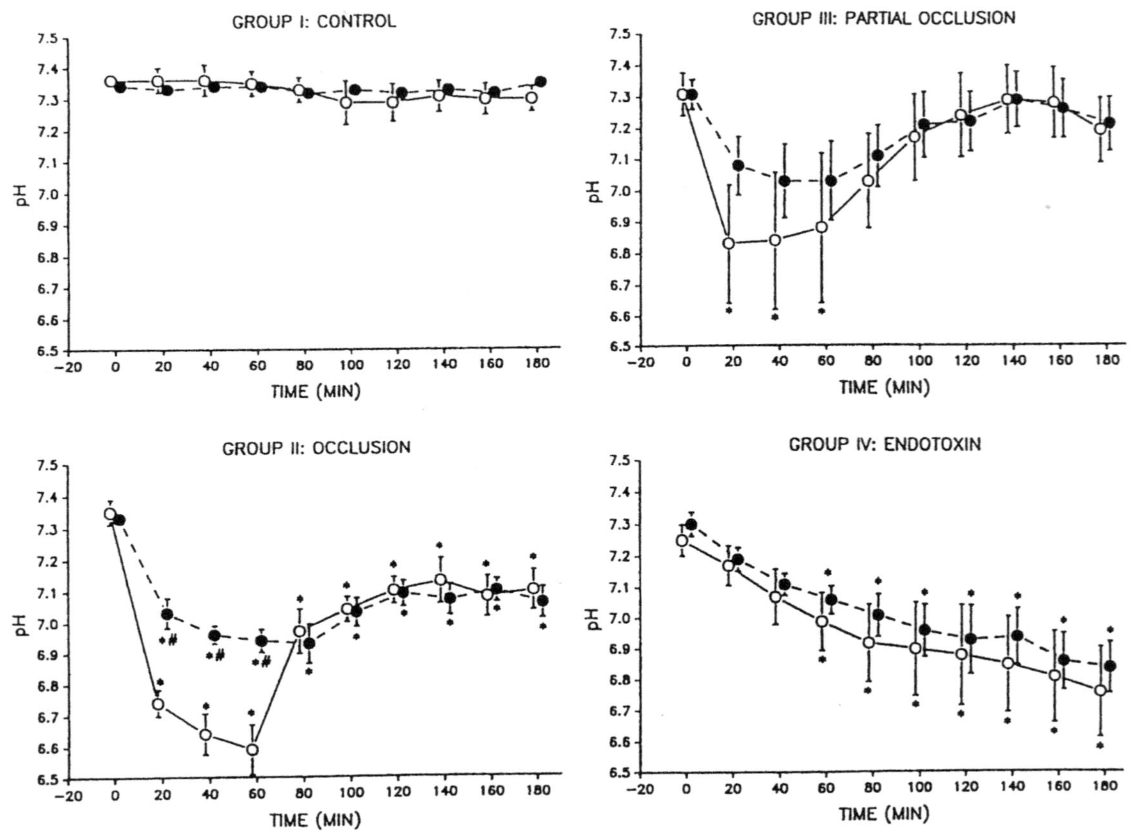

Fig. 4. pHi measured by microelectrodes (*open circles*) and estimated tonometrically (*filled circles*) in pigs. *Group I*, saline controls; *group II*, complete mesenteric occlusion from 0 to 60 min; *group III*, partial mesenteric occlusion from 0 to 60 min; *groups IV – IV*, *E. coli* endotoxin infused from 0 to 20 min. * $p<0.5$ vs. baseline; # $p<0.05$ between methods. (From [10])

nificance (Fig. 4). Thus, local bicarbonate consumption is expected to induce only a slight overestimation of tonometrically measured pHi in patients with gut dysoxia. This error is not measurable in clinical practice, and arterial bicarbonate should therefore be accepted as an estimator of tissue bicarbonate.

General Questions

What Is a Normal pHi?

There is no defined limit for normal pHi values. The majority of studies use values of 7.35 or 7.32 as lower normal limits. The reasons for chosing these values were unclear. Fiddian-Green [30] suggested that these values are chosen because they are approximately one or two standard deviations from the mean (7.39 ± 0.03) pHi of 14 healthy volunteers in the study of Heard et al. [15]. No calibration studies aimed at establishing the most discriminative cutoff value have been published to date. Also, it is not clear what pHi value should be used as a therapeutic goal. It has even been suggested that moderate local acidosis might protect tissue cells from damage induced by ischemia, and that bicarbonate administration can have deleterious effects [30]. In a study using pHi as a therapeutic index Gutierrez el al. [9] observed that therapeutic interventions aimed at maintaining pHi in values over 7.35 increased survival in the subgroup of patients with normal (over 7.35) pHi at admission. This value was chosen from a previous study on the prognostic value of pHi [4].

Reliability and Variability: A Statistical Model for Reliability

What does the statistical concept of reliability mean [34]? No matter what the variable and no matter how it is obtained (by physical examination or by interviewing or by laboratory assay), it is measured unrealibly in the sense that, when the individual is measured again under similar conditions, the second value would differ to some extent from the first. Imagine a subject being repeatedly measured on the variable X of interest under conditions as close to uniform as possible, and let T denote the mean of the hypothetical replicate measurements on him. T is referred to in psychometrics as the subject's "true score," but less association-laden expressions include "error free score," "steady-state value," and "signal". A single measurement X may differ from T for any number of reasons: random coding of errors, inherent lability of the characteristic, or imperfect calibration of a measuring device. If e represents the difference between a single observation on a subject and its underlying mean value T, the classical linear model for an observed score is obtained: $X = T + e$. In a population of subjects the error-free score T varies about some mean value m with a variance of σ_T^2. For a single subject the random error e varies about a mean of zero. Under the assumption that the distribution of the errors is independent of the value of T, e has a variance

Table 3. Layout of data from a simple reproducibility replication reliability study

Subject	Number of measurements	Mean	Variance
1	k_1	X_l	s_l^2
i	k_1	X_1	s_1^2
N	k_n	X_n	s_n^2
Total	K	X	S^2

of σ_e^2 regardless of the value of T, and therefore the variance of X is: $\sigma_X^2 = \sigma_T^2 + \sigma_e^2$. This expresses the phenomenon that there are two components of the variability among a series of measurements on different subjects: that among their steady-state values plus the variability of the random errors.

A single quantity that conveniently expresses the relative magnitude of the two components of the variance of X is the intraclass correlation coefficient of reliability (R). $R = (\sigma_T^2)/\sigma_T^2 + \sigma_e^2)$. As σ_e^2/σ_T^2 decreases, error constitutes a decreasing portion of the observed values; reliability therefore increases, and R approaches its maximum value, unity. Conversely, as σ_e^2/σ_T^2 increases, error constitutes an increasing portion of the observed values; reliability therefore decreases, and R approaches its minimum value, zero. R is directly interpretable as a proportion of variance. It is the proportion of the variance of an observation due to subject-to-subject variability in error-free scores.

Reliability studies allow the one to evaluate the reproductibility of the measurements obtained by a given method. Reliability studies can be carried out by performing k repeated measurement on N subjects. The results of this simple kind of reliability or reproducibility study are summarized as in Table 3. For a typical subject, say the ith, k_i is the number of replicate measurements on him or her, X_i is the mean of the k_i measurements, and s_i^2 is their variance. Therefore, $X_{i1}, X_{i2}, \ldots X_{iki}$, represents the k_i measurements on subject i:

$$\bar{X}_i = \frac{1}{k_i} \sum_{j=1}^{k_i} X_{ij}$$

and:

$$s_i^2 = \frac{1}{k_i - 1} \sum_{j=1}^{k_i} (X_{ij} - \bar{X}_i)^2$$

In the final row of Table 3 K is the total number of measurements:

$$K = \sum_{i=1}^{N} k_i$$

Table 4. Results of a simple replication reliability study (5 measurements per patient)

Patient	Mean	Variance
1	7.453866	0.001519
2	7.321664	0.001386
3	7.446022	0.002905
4	7.329898	0.012432
5	7.420492	0.01291
6	7.331066	0.000125
7	7.324027	0.000497
8	7.326485	0.006853
9	7.282716	0.000404
10	7.297408	0.00017
11	7.430105	0.000302

X is the overall mean:

$$\bar{X} = \frac{1}{K} \sum_i \sum_j X_{ij} = \frac{1}{K} \sum_i k_i \bar{X}_i;$$

and S^2 is the overall variance:

$$S^2 = \frac{1}{k-1} \sum_i \sum_j (X_{ij} - \bar{X})^2$$

To assess the reliability of individual pHi measurements five consecutive determinations of pHi obtained in each of 11 patients were analyzed. The measurements were obtained with an equilibration period of 2 h for the patients included as a control group in a study evaluating the effect of ranitidine on pHi determinations in critically ill patients (Table 4). Variability between the means is appreciably greater than the average of the several within-patient variances, suggesting that *within*-patient variability (σ_e^2) is much smaller than *between*-patient variability (which is informative about σ_T^2). The analysis of variance provides a quantitative rather than qualitative description of the two components of variability. The statistical calculations for analysis of variance on data such as those in Table 3 appear on the left in Table 5 and the quantities obtained by applying these to the data in Table 4 appear on the right.

The constant k_O that appears in the expected value of the between-patient mean square is equal to:

$$k_0 = \bar{k} - \frac{s_k^2}{N\bar{k}}$$

Table 5. Analysis of variance for the results of a simple replication reliability study

	General statistical expression				Data in Table 4		
Source of variation	df	SS	MS	E(MS)	df	SS	MS
Between patients	$N-1$	$\sum k_i(\bar{X}_i-\bar{X})^2$	BMS	$\sigma_e^2+k_0\sigma_T^2$	10	0.20152	0.020152
Within patients	$K-N$	$\sum (k_i-1)s^2$	WMS	σ_e^2	44	0.03950	0.000897
Total	$K-1$	$(K-1)S^2$	–	–	54	0.24102	-

df, Degrees of freedom; SS, sum of squares; MS, mean square (quotient of the sum of squares divided by the degrees of freedom); E(MS) expected mean square (the underlying statistical quantity estimated by MS); BMS, between-patient mean square; WMS, within-patient mean square.

where k and s^{2k} are the mean and variance of the numbers of replicate measurements:

$$\bar{k} = \frac{K}{N}$$

and:

$$s_k^2 = \frac{1}{N-1} \sum_i (k_i - \bar{k})^2$$

If, as in our study these numbers are constant (i.e., if $k_1 = \ldots = k_n = k$) then $k = k$, $s_k^2 = 0$, y $k_O = k$. The within-patient mean square (WMS) is an unbiased estimator of σ_e^2, the component of variance due to random error. It is therefore convenient for current purposes to let s_e^2 designate WMS. The quantity:

$$s_T^2 = \frac{\text{BMS} - \text{WMS}}{K_0}$$

is an unbiased estimator of σ_T^2, the component of the variance error-free variability among subjects. An estimator of the intraclass correlation coefficient is then:

$$R = \frac{s_T^2}{s_T^2 + s_e^2} = \frac{\text{BMS} - \text{WMS}}{\text{BMS} + (K_0 - 1)\,\text{WMS}}$$

Although R is the ratio of two unbiased estimators, it itself is a slightly biased estimator of the parameter R. The bias becomes negligible as N increases. For the present data the estimated component of variance due to random measurement error is:

$$s_e^2 = \text{WMS} = 0.000897$$

Because the constant number of replicate observations per patient is $k=5$, therefore $k_0=5$, and the estimated component of variance due to variability of error-free scores is:

$$s_T^2 = \frac{\text{BMS} - \text{WMS}}{k} = \frac{0.020152 - 0.000897}{5} = 0.003851$$

The estimated intraclass correlation coefficient for the variable being analyzed is:

$$R = \frac{s_T^2}{s_T^2 + s_T^2} = \frac{0.003851}{0.003851 + 0.000897} = 0.81107$$

a value indicating fairly high reliability. The square root of WMS, s_e, is referred to as the standard error of measurement. One of its most important uses is the following. For a single subject the error-free score T may be considered an unknown parameter. If a single observed measurement X were available on that subject, an approximate 95% confidence interval for his or her underlying T, would be $X \pm 2s_e$. In our case:

$$s_e = \sqrt{0.000897} = 0.0299$$

and thus, the corresponding limits for a 95% confidence interval for T are between the margins of $X \pm 0.06$.

These values indicate that pHi measurement by tonometry has a fairly high reliability. Reliability can be further improved by using repeated measurements and rejecting aberrant determination, two methods that are routinely applied to biological measurements, for example in measuring cardiac output. The indexes obtained are, however, relative to the measurements obtained in the study conditions, with an equilibration period of 2 h. Data on reliability of pHi measurements with shorter equilibration periods are not available.

Conclusion

pHi measurement by tonometry is a fairly reliable technique, with reproducible results. The necessity of ranitidine administration to measure pHi in critically ill patients remains to be determined. Sucralfate does not seem to interfere with pHi monitoring, but CO_2-producing antiacids should be avoided. There are no published data about the effect of enteral nutrition on pHi, but enteral nutrition should probably be stopped before measuring pHi. Corrected indexes such as pHi gradient or standarized pHi may be used in patients with acid-base disturbances although their clinical value requires elucidation.

Specific calibration is necessary for measuring CO_2 in saline with certain gas analyzers. The next goal to achieve is developing a continuous pHi monitoring device. This would overcome most of the current technical problems of tonometry. The application and interpretation of pHi measurements, however, remain to be defined.

References

1. Gutierrez G, Bismar H, Dantzker D, Silva N (1992) Comparison of gastric intramucosal pH with measures of oxigen transport and consumption in critically ill patients. Crit Care Med 20:451–457
2. Fiddian-Green RG, Gantz NM (1987) Transient episodes of sigmoid ischemia and their relation to infection from intestinal organism after abdominal aortic operations. Crit Care Med 15:835–839
3. Fiddian-Green RG, Baker S (1987) Predictive value of the stomach wall pH for complications after cardiac operations: comparison with other monitoring. Crit Care Med 15:153–156
4. Doglio GR, Pusajo JF, Egurrola MA et al (1991) Gastric mucosal pH as a prognostic index of mortality in critically ill patients. Crit Care Med 19:1037–1040
5. Maynard N, Bihari D, Beale R et al (1993) Assessment of splachnic oxygenation by gastric tonometry in patients with acute circulatory failure. JAMA 270:1203–1210
6. Gys T, Hubens A, Neels H, Lauwers LF, Peeters R (1988) Prognostic value of gastric intramural pH in surgical intensive care patients. Crit Care Med 16:1222–1224
7. Marik PE (1993) Gastric intramucosal pH. A better predictor of multiorgan disfunction syndrome and death than oxigen-derived variables in patients with sepsis. Chest 104:225–229
8. Landow L, Phillips DA, Heard S, Prevost D, Vandersalm TJ, Fink MP (1991) Gastric tonometry and venous oximetry in cardiac surgery patients. Crit Care Med 19:1226–1233
9. Gutierrez G, Palizas F, Doglio G et al (1992) Gastric intramucosal pH as therapeutic index of tissue oxygenation in critically ill patients. Lancet 339:195–199
10. Antonsson JA, Boyle CC III, Kruithoff KL et al (1990) Validation of tonometric measurement of gut intramural pH during endotoxinemia and mesenteric occlusion in pigs. Am J Physiol 259:G519–G523
11. Groeneveld ABJ, Kolkman JJ (1994) Splanchnic tonometry: a review of physiology, metodology and clinical applications. J Crit Care 9:198–210
12. Schlichtig R, Bowles SA (1994) Distinguishing between aerobic and anaerobic appearance of dissolved CO_2 in intestine during low flow. J Appl Physiol 76(6):2443–2451
13. Fiddian-Green RG, Pittenger G, Whitehouse WM (1982) Back- diffusion of CO_2 and its influence on the intramural pH in gastric mucosa. J Surg Res 33:39–48
14. Kolkman JJ, Groeneveld ABJ, Meuwissen SGM (1994) The effect of ranitidine on basal and bicarbonate enhaced intragastric PCO2:a tonometric study. Gut 35:737–741
15. Heard SO, Helsmoortel CM, Kent JC et al (1991) Gastric tonometry in healthy volunteers: effect of ranitidine on calculated intramural pH. Crit Care Med 19:271–274
16. Baigorri F, Calvet X, Duarte M et al (1994) Effect of ranitidine treatment in gastric intramucosal pH determination in critical patients. Intensive Care Med 20:S2
17. Maynard N, Atkinson S, Mason R, Smithies M, Bihari D (1994) Influence of intravenous ranitidine on gastric intramucosal pH in critically ill patients. Crit Care Med 22:A79
18. Stannard A, Hutchinson A, Morris DL, Byrne A (1988) Gastric exocrine "failure" in critically ill patients: incidence and associated features. Br Med J 296:155–156

19. Hunt RH (1992) Acid, bacteria and the mucosal barrier: aspiration or translocation as the cause of pneumonia? Eur J Gastroenterol Hepatol 4:881–884
20. Higgins D, Mythen MG, Webb AR (1994) Low intramucosal pH is associated with failure to acidify the gastric lumen in response to pentagastrin. Intensive Care Med 20:105–108
21. Driks MR, Craven DE, Celli BR et al (1987) Nosocomial pneumonia in intubated patients given sucralfate as compared with antiacids or histamine type 2 blockers. The role of gastric colonization. N Engl J Med 317:1376–1382
22. Du Moulin GC, Paterson DG, Hedley-Whyte J, Lisbon A (1982) Aspiration of gastric bacteria in antiacid-treated patients: a frequent cause of postoperative colonisation of the airway. Lancet 1:242–245
23. Soll AH (1993) Gastric, duodenal and stress ulcer. In: Sleisenger MH, Fordtran JS (eds) Gastrointestinal disease: pathophysiology, diagnosis, management, 5th edn. Saunders, Philadelphia, pp 581–678
24. Calvet X, Baigorri F, Duarte M et al (1994) Effect of sucralfate treatment in gastric intramucosal pH determinations in critical patients. Intensive Care Med 20:S5
25. Odes HS, Hogan DL, Steinbach JH et al (1992) Measurement of gastric bicarbonate secretion in the human stomach: different methods produce discordant results. Scand J Gastroenterol 27:829–836
26. Rune SJ, Henriksen FW (1969) Carbon dioxide tensions in the proximal part of canine gastrointestinal tract. Gastroenterology 56:758–762
27. Stevens MH, Thirlby RC, Feldman M (1987) Mechanisms for high PCO_2 in gastric juice: roles of bicarbonate secretion and CO_2 diffusion. Am J Physiol 253:G527–G530
28. Salzman AL, Strong KE Wang H, Wollert PS, Vandermeer TJ, Fink MP (1994) Intraluminal "balloonless" air tonometry: a new method for determination of gastrointestinal mucosa carbon dioxide tension. Crit Care Med 22:126–134
29. Fiddian-Green RG (1993) The pH gap: a meassurement for establishing the presence or absence of intestinal mucosal ischemia independently of disturbances in sistemic acid base balance. Eur Surg Res 25 [Suppl 1]:25–26
30. Fiddian-Green RG (1994) Management of gastric intramucosal acidosis. In: Vincent JL (ed) Yearbook of intensive care and emergency medicine. Springer, Berlin Heidelberg New York, pp 217–224
31. Riddington D, Venkatesh B, Clutton-Brock T, Bion J (1992) Potential hazards in estimation of gastric intramucosal pH. Lancet 340:547
32. Riddington D, Venkatesh KB, Clutton-Brock T, Bion J (1994) Measuring carbon dioxide tension in saline and alternative solutions: quantification of bias and precision in two blood gas analyzers. Crit Care Med 22:96–100
33. Knichwitz G, Mertees N, Kuhmann M (1994) Improvement of intramucosal pH measurements with a phosphate-buffered solution. Intensive Care Med 20:S137
34. Fleiss JL (1986) Reliability of meassurement. In: Fleiss JL (ed) The design and analysis of clinical experiments. Wiley, New York, pp 1–31

Gut Permeability

T. Hassanein, D. H. Van Thiel, and A. Gurakar

Introduction

The gut is comprised of a series of unique organs that absorb nutrients, electrolytes, and water and simultaneously secrete other electrolytes and water while it also retards the progress of ingested food down its length. The gut is continuously challenged with contents that vary both in their volume and composition. It limits the absorption of large, potentially harmful molecules such as antigens and toxins while absorbing smaller nontoxic substances with either a nutritional or caloric value [1]. The intestinal mucosa acts as a protective barrier between the external and internal milieu through which molecules of various size cross either by active or passive transport mechanisms which vary as a function of the anatomical region of the gut in question [2, 3]. This barrier is armored by intraluminal factors [4, 5], interepithelial factors [3], and extraintestinal factors [6]. The structure of the surface epithelium with its local mucosal immune system is the main pillar of the barrier [7]. Intraluminally, gastric acid, pancreatic and intestinal enzymes, intestinal motility, and the mucous coat covering the microvilli combined with secretory antibodies and colonization-resisting flora each strengthen the intestinal barrier and limit the proliferation and systemic dissemination of enteric pathogens. This system is supported by third-line defenses represented by the lymphoid system of the gut, liver, spleen, and systemic immunoglobulins which clear translocated pathogens and particles.

Characteristics of the Small Intestinal Epithelium

The epithelial cells of the small intestine have their apical surfaces facing the intestinal lumen and their basolateral surfaces defined by the intercellular junctions and the intestinal basement membrane (Fig. 1). The basolateral membrane has three types of junctions [8]:

- Tight junctions located at the apical pole of the cell: These are fused regions of adjacent lateral cell membranes that occlude the intercellular space. Tight junctions contribute indirectly to the transepithelial permeability barrier by regulating the diffusion of ions and neutral molecules through the intracellular space. They also participate in the separation of

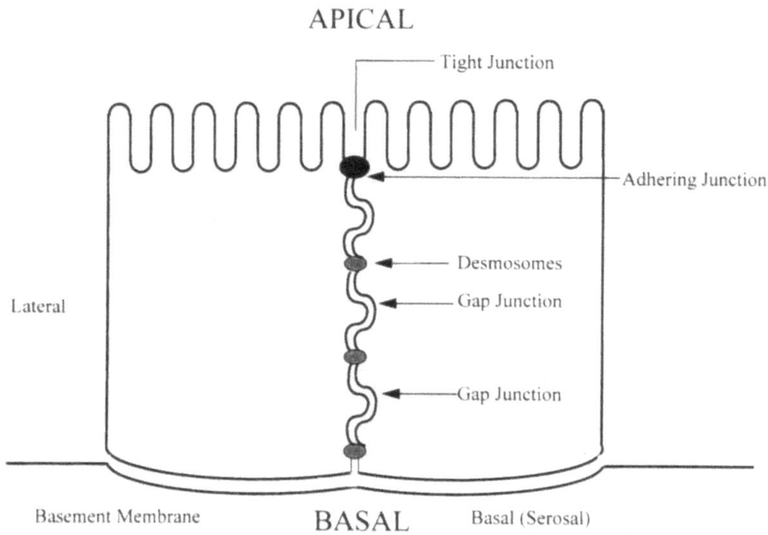

Fig. 1. Junctions of the surface epithelium

the epithelial cell surface membrane into apical and basolateral domains. The "tightness" of these junctions varies in the different segments of the gut. In the jejunum tight junctions are more leaky than in the distal small bowel, colon, or rectum. The distribution of tight junctions along a villus is not constant. Tight junctions are fewer in number at the tip of each villus. The movement of solutes and ions across or through the pericellular space is by simple diffusion. Although passive, it is highly regulated. Consequently the permeability of the tight junctions is a major factor in determining the rate and character of the passive fluxes across the intestinal epithelium. The maintenance of the integrity of tight intercellular junctions between adjacent enterocytes is a dynamic process, actively controlled by adenosine triphosphate-dependent intracellular mechanisms and by the cytoskeleton of the epithelial cell [9, 10].

- Adhering junctions: These link the cell membrane and the cytoskeletal elements between adjacent cells to allow for the distribution of mechanical forces.
- Gap junctions: These are actual spaces between adjacent cells which permit the passage of ions and small molecules.

Epithelia are either tight (e.g., urinary bladder) or leaky (kidney and small intestine) depending on their individual permeability characteristics [11]. Leaky epithelia are characterized by low-resistance pericellular pathways. Small molecules having a radius of 4 Å or less and are water soluble, such as mannitol and mannose, pass through numerous small water-filled pores within the cell membranes of enterocytes [12]. Larger molecules (lactulose and cellobiose) can only pass through the pericellular channels. Ions and solutes transported passively across the epithelium do so as the result of either diffusion or convection (solvent drag) determined by the differences in the concentrations of the solute at two sites and the electrical potential across the mucosal membrane [13]. Passive transport can occur through cells (transcellular) or through the tight junctions (pericellular) and does not require energy.

In contrast, active transport across an epithelial surface is always transcellular and requires energy. The movement of ions and solutes into a cell (transcellular) through the lipid bilayer of the cell membrane is controlled by channels, pumps, and carriers, which are specialized protein components of the cell membrane. Channels are specific for the passage of particular ions which pass through them by electrodiffusion. The opening and closing of specific channels is controlled by a wide variety of different signals. Carriers are an even more specific means for the transport of certain solutes and ions. They are involved in both facilitated diffusion and active transport of ions. They can be saturated or inhibited. Finally, pumps mediate primary active transport mechanisms against an electrochemical gradient to maintain cell volume and osmotic balance [1].

The intestinal epithelial surface is a relatively impermeable barrier to the absorption of large water soluble molecules. Absorption of such molecules varies inversely with their molecular size and weight such that less than 1% of the substances with a molecular weight of more than 200 g/mol are absorbed [12]. Thus most molecules with a cross-sectional diameter greater than 4 Å are absorbed only to a limited extent [13–15]. However, smaller molecules can be absorbed more easily. The barrier function of the mucosa depends primarily on an intact surface epithelial cell layer and on the tightness of the tight junctions between adjacent epithelial cells. The ability of the mucosa to selectively absorb needed nutrients and exclude harmful molecules has been referred to as "selective permeability" [12, 16].

Various immunological mechanisms support the mucosal barriers. Gastrointestinal lymphoid tissues such as Peyer's patches, lymphocytes, macrophages and IgA-producing plasma cells provide a defense mechanism against translocation of harmful intraluminal contents, while Kupffer's cells and endothelial cells of the liver serve as a second line of defense system against bacteria and their toxins, preventing their access to the systemic circulation [3, 17, 18]. The toxins which gain access to the circulation, can be cleared or detoxified by certain serum proteins such as lipopolysaccharide binding protein, high density lipoprotein and others. It is also endocytosed by endothelial and Kupffer cells [19]. Endotoxins are lipopolysaccharide constituents of the outer membranes of gram-negative bacteria. When they gain access to the systemic circulation, they induce a variety of biological effects including the induction of intravascular coagulation [20], hemodynamic instability [21], a host of metabolic derangements [22] that contribute to damage to the vascular endothelium [23], cholestasis, suppression of cellular immunity, and increased muscle protein degradation [24, 25].

Intestinal permeability is affected by a wide variety of intestinal and systemic disorders such as hemorrhage, shock [26], thermal injury [27], endotoxemia, trauma, intestinal obstruction, and malnutrition [28]. Obviously any gross disruption of the epithelial cell layer results in a perturbation of epithelial barrier function and permeability. The integrity of the permeability barrier depends in large degree upon gut perfusion. Low perfusion states, which occur in different situation, result in gut mucosal ischemia [29, 30]. Because of the peculiar anatomical features of the gut mucosa, perfusion of the splanchnic bed decreases dramatically and out of proportion to the reduction in total cardiac output [31, 32]. In situations in which the intestinal barrier is grossly disrupted, intestinal bacteria, particularly gram-negative aerobes and toxins, can migrate across the epithelium to reach mesenteric lymph nodes and the portal and systemic circulations resulting in bacterial sepsis and/or endotoxemia. This process is called translocation [2, 4, 33]. Bacterial translocation is defined as the passage of viable bacteria from the gastrointestinal tract to extraintestinal sites such as mesenteric lymph nodes, spleen, liver, kidney, and systemic blood. It occurs continuously at low levels via direct transmucosal migration across the bowel wall to mesenteric lymph nodes, liver, and spleen by lymphatic or vascular channels. Translocation of minute amounts of endotoxin into the portal circulation is an important physiological phenomenon which enhances the activity of the reticuloendothelial system [2]. However, systemic endotoxemia results in organ dysfunction. Under normal circumstances only small amounts of endotoxin are absorbed from the gastrointestinal tract. These are normally detoxified by the liver [34]. However, with greater degrees of translocation, systemic endotoxemia occurs and it overwhelms the reticuloendothelial system, causing a decrease in its functions [34, 35]. Endotoxin can also reach the systemic circulation via the peritoneal cavity after translocating across the bowel wall [39].

Bacterial translocation is controlled by multiple factors such as colonizing resistance and gut membrane barrier [36]. Thus, enhanced bacterial translocation can result from: (a) an increased permeability of the intestinal mucosa barrier as seen in hemorrhagic shock and sepsis, (b) a decrease in host defense mechanisms as occurs in protein malnutrition and immunosuppression, or (c) disruption of the normal ecological system in the intestine allowing bacterial overgrowth and intestinal stasis as occurs in cases of intestinal obstruction.

Membrane Permeability

Host immune defenses are considered an integral component in the processes that limit bacterial translocation. Bacteria translocated through the intestinal mucosal barrier reach the mesenteric lymph nodes (MLN) before they reach the liver, spleen, peritoneal cavity, or blood [34]. Most translocating bacteria are killed in transit or in the MLN by the mucosal, cell-mediated, and humoral immune effector mechanisms of the host. Secretory immunoglobulins such as IgA inhibit the adherence of bacteria to the mucosal epithelial cell. Macrophages present in the lamina propria and lymphoid organs act as a second line of defense against translocated bacteria while serum immunoglobulins opsonize translocated bacteria in the blood stream [37].

Membrane permeability is a property which describes the free diffusion of a specific solute across a membrane over a given period of time [38]. Permeability measurements provide a more sensitive indication of mucosal integrity than is achievable with either carrier-mediated absorption or histopathological studies. The principal two factors affecting membrane permeability are: (a) the intrinsic structure of the membrane and (b) the physiochemical properties of the solute in question.

Detecting subtle changes in the permeability of the intestinal epithelium in various disease processes provides important insights into the mechanisms and consequences of the many disease processes that affect the small intestine either directly or indirectly and can be accomplished using a variety of model probes. Gut permeability can be estimated by measuring the rate of flux of a test or model substance through a defined region of the intestinal membrane. Mucosal damage decreases the absorptive area of the intestine, a factor which determines the permeation of small molecules and increases the number of larger pores which enable the permeation of larger molecules [39].

The permeability of a given substance is determined by its molecular size in relation to the size of the water-filled pores within the membrane. Hence different sized probes are typically used to determine intestinal permeability. The normal intestine discriminates between various permeability probes according to their cross-sectional diameters and their lipophilic or hydrophilic characteristics [40]. Once across the mucosa, these probes gain access to the systemic circulation and are excreted by the kidney and can be recovered and quantitated in urine to determine noninvasively the permeability of an intestinal membrane. Increased permeability to probes such as lactulose and ^{51}Cr-labeled ethylenediamine tetracetate (^{51}Cr-EDTA) are correlated with increased permeation of macromolecules and chemotactic bacterial peptides [41–43].

Certain noninvasive permeability assessment techniques are so simple and reliable that they can be used to screen individuals for the presence of small bowel disease. In 1974 Menzies and coworkers introduced the use of nonhydrolyzable di- and trisaccharides as test substances for quantitating intestinal permeability (e.g., lactulose, l-rhamnose) [44]. Chadwick and coworkers introduced the use of polyethylene glycols (PEG^{400}) in 1977 [45]. In 1983 ^{51}Cr-EDTA was introduced by Bjamason et al. to access the barrier function of the small intestine [46].

All of these test substances fulfill to a variable degree the theoretical criteria that define an ideal permeability probe [40]. These include that the model substance or probe be physicochemically inert, water soluble, nontoxic, easily measurable in biological fluids, inexpensive, not degradable or metaboliz-

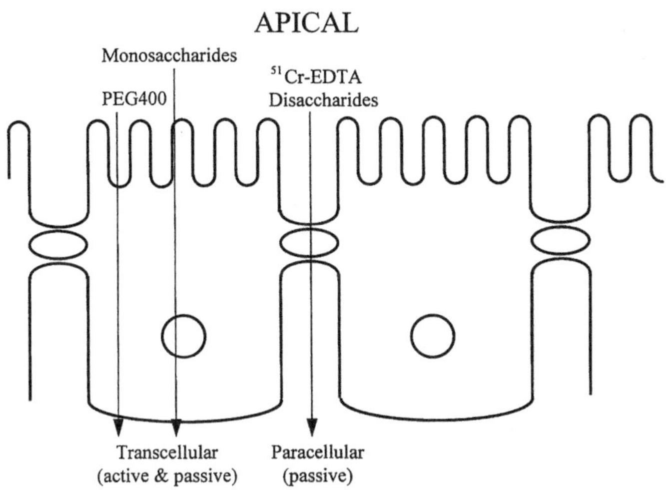

Fig. 2. Permeability pathways of the different probes used in permeability studies. *PEG400,* polyethylene glycols; *^{51}Cr-EDTA,* ^{51}Cr-labeled ethylenediamine tetracetate

able before, during, or after intestinal absorption; that it have first order kinetics of absorption, a precise and accurate method for detection, a urinary excretion rate that is rapid and complete; and the potential for use in a noninvasive test procedure.

The potentially near-ideal probes or model substances useful in assessing intestinal permeability currently in clinical use include sugars, PEG^{400}, and 51Cr-EDTA.

Sugars. These include disaccharides (lactulose, cellobiose) and monosaccharides (L-rhamnose, mannitol, mannose). Lactulose is a globular sugar with a MW of 342 g/mol and a cross-sectional diameter of 9.5 Å. L-rhamnose is also a globular sugar with a MW 164 g/mol and diameter of 8.3 Å. Mannitol is a linear sugar with a MW 182 g/mol and diameter of 6.7 Å. Because of their size disaccharides such as lactulose are absorbed minimally usually less than 0.5%, presumably paracellularly through the tight junctions or through interruptions in the epithelial surface at the extrusion zones on villus tips (Fig. 2). The precise mechanism of lactulose transport across the intestinal epithelial barrier, however, is not known. Moreover, the possibility that several routes are involved is real. Its transport may be either passive diffusion along a concentration gradient or by endocytosis [47]. On the other hand, monosaccharides, which are smaller molecules, cross the intestine primarily by way of aqueous pores in the enterocyte brush border membrane. Thus 20 to 100 times more monosaccharides can be absorbed under normal circumstances. They also can pass in minute amounts through intercellular junctions. Thus the amount of lactulose absorbed depends primarily on the integrity of the small intestine epithelium and tight junctions while mannitol and L-rhamnose absorption depend more on total small intestinal surface area.

Assessment of differential absorption is an absorption technique in which appropriate paired sugar markers are ingested simultaneously and detected in the urine collected over a certain period of time. The results are expressed as a ratio of their urinary excretions. This method overcomes the effects of confounding variables such as gastric emptying, intestinal dilution and transit, systemic distribution and metabolism, renal clearance, and adequacy of urine collection. The substances commonly used are PEG^{400} and ^{51}Cr-EDTA.

PEG^{400}. This is a linear polymer with a cross-sectional diameter of 5.3 Å. PEG^{400} is a commercially available mixture of at least nine distinct PEG polymers. The molecular weight of PEG^{400} preparations varies from 238 to 590, with a mean of 400, and the percentage of individual polymers present in a given preparation follows a gaussian distribution about the mean. PEG^{400} does not undergo either tissue or bacterial degradation, and it is excreted rapidly into urine. The differential pattern of urinary recovery of various PEG^{400} subspecies is attributed to differential intestinal permeation [40]. PEG^{400} is absorbed predominantly through the lipid-rich, hydrophobic domains of the brush border (Fig. 2). It is also transported passively through tight junctions, being undiluted by a solvent drag. Its passage through this latter site is not quantitatively important in its overall absorption [38, 48].

^{51}Cr-EDTA. This is normally a nonabsorbable substance; in disease states it passes through the tight cellular junctions between adjacent enterocytes (Fig. 2). ^{51}Cr-EDTA is closely related in size, shape and permeability characteristics to lactulose. Both probes have an increased permeability through inflamed mucosa [49].

Basic Permeability Methods

Sugars

The usual test is performed using a mixture of disaccharides (5–10 g) and a monosaccharide (0.5–2 g) given orally after an overnight fast. Urine is collected for 5–6 h. Chromatography or enzymatic analysis is used to determine the quantity of each component of interest in the urine collected. Results are expressed as excretion ratios of the disaccharide to monosaccharides (the ratio is an index of the function of the permeability pathways for each). Theoretically the calculated ratio is independent of problems associated with an incomplete urine collection, gastric and intestinal transit, intestinal dilutional of the probe, and altered renal function since these potentially confounding factors affect both markers to a similar extent. The disaccharides, however, are given in amounts which are osmotically active and can cause intestinal hurry, thereby bypassing proximal areas of interest and also causing abdominal distention, diarrhea, and flatulence [47, 49–52].

Polyethylene Glycol

PEG^{400} tests are performed after the ingestion of 5–10 g PEG after an overnight fast. Urine is collected for 5–6 h and is assayed for its PEG content by high performance liquid chromatography. PEG^{400} is a mixture of seven or eight PEG polymers ranging in MW from 200 to 500. The idea in using PEG is to determine the permeability of various PEG polymers over the range of MW defined by the specific PEG preparation being used by comparing the ratios of the different polymers absorbed and subsequently excreted and measured in the urine. A peculiarity of this technique is that the percentage urinary recovery of the polymers is not reduced as a result of increasing polymer size. PEG^{400} is an important probe of intestinal permeability because it is the only permeability probe in current use which crosses the intestine via a lipophilic route [48, 53, 54].

^{51}Cr-EDTA

After an overnight fast the subject ingests 100 mCi ^{51}Cr-EDTA (3.7 MBq) and 200–300 ml water. Urine is collected for 24 h in two portions. The first 6-h collection reflects upper gastrointestinal permeability. The latter 18-h collection reflects distal small bowel and colonic permeability. ^{51}Cr-EDTA is measured in the urine using a gamma counter.

The ^{51}Cr-EDTA method differs from the disaccharide method in three important ways:

(a) ^{51}Cr-EDTA is not degraded by bacteria, as are disaccharides;
(b) it is absorbed in part by the colon (the amount absorbed by the colon cannot be determined precisely) especially in small bowel diseases where secretory diarrhea enhances the load delivered to the colon.
(c) ^{51}Cr-EDTA is administered in trace amounts (<0.6 mmol) which do not interfere with normal intestinal physiology, while disaccharides are administered in physiological amounts.

Although ^{51}Cr-EDTA is a radioactive substance, it does not pose a clinical hazard when used in permeability tests because very little is actually absorbed, and even less is retained by the patient. The whole-body radiation exposure achieved is calculated to be 0.001 mGy. This technique is susceptible to alterations as a result of the ingestion of alcohol and/or nonsteroidal anti-inflammatory drugs. The major disadvantage of the EDTA method is that it requires that a complete 24-h urine collection to be obtained [49, 55].

Intestinal Permeability of the Gut in Different Clinical Situations

Hemorrhagic Shock

Hemorrhagic shock is a common clinical problem associated with many clinical situations such as trauma and gastrointestinal bleeding in patients with portal hypertension. Most patients survive the initial insult, only to develop sepsis and multiorgan failure days or weeks later. Most of the sepsis occurs due to endogenous bacteria which have translocated from the intestinal tract. Studies in rodents have revealed bacterial translocation occurring as early as 2 h after hemorrhagic shock during the acute phase of ischemic mucosal damage [29, 30]. In baboons and pigs hemorrhagic shock causes bacterial translocation to mesenteric lymph nodes as a late event (>6–8.5 h after hemorrhage) at a time when yet bacteria are not detectable in the portal blood, liver, or spleen [56]. However, these models are not analogous to humans. Many studies in man have reported the presence of enteric bacteria in blood within hours of hemorrhagic shock [57]. In trauma patients Peitzman et al. found that bacterial translocation to MLNs is not common in acutely injured patients [58].

However, for bacteria to reach the systemic circulation they do not have to pass through MLNs. Other routes exist, such as direct passage through the enterocyte (intracellular pathway) and passage between intestinal epithelial cells (intercellular) and through the muscle layer of the bowel wall to the peritoneal cavity. No histological studies of the gut mucosa have reported any evidence of clinically significant mucosal cell damage early after shock. In rats Parks et al. reported that shock leads to an ischemic reperfusion injury of the intestinal mucosa. They implicate oxygen free radicals as mediators of this injury [59]. Deitch et al. showed that the translocation in rats is reduced in animals given allopurinol prior to shock [60, 61]. They suggest that oxygen free radicals generated during ischemia/reperfusion are the major cause of the damage to the mucosal barrier that leads to subsequent bacterial translocation. A similar sequence has been reported to occur in patients with thermal injury [62].

Mesenteric ischemia for 3–4 h does not cause mucosal damage; however, during reperfusion mucosal damage begins to be histologically detectable. This mucosal injury is related to the presence of reactive oxygen metabolites (superoxide radical or hydrogen peroxide and hydroxyl radical). Normally the gut is protected against injury due to reactive oxygen metabolites because of its content of free radical scavengers (superoxide dismutase, catalase, and glutathione peroxidase). During ischemia ATP is hydrolyzed to adenosine which subsequently is converted to hypoxanthine by xanthene oxidase, which increases during periods of ischemia. During reperfusion this ischemia induced xanthene oxidase catalyses the conversion of hypoxanthine to xanthine and uric acid. These reactions result in the formation of hydrogen peroxide and superoxide radical ions. These ions produce enzyme inactivation of membrane transport proteins and peroxidation of unsaturated bonds in membrane lipids causing tissue damage. This sequence of events has been termed the "xanthene oxidase hypothesis" of reperfusion injury [4]. It results in intestinal barrier dysfunction and alterations in mucosal permeability leading to bacterial and toxin translocation [63].

Thermal Injury

Thermal injury and burns result in gut barrier failure and increased intestinal permeability. This failure is transient, and restoration of normal function occurs within days unless the burned area gets infected. In humans intestinal permeability to lactulose is significantly longer in infected burned patients than in those whose clinical courses are not complicated by infection [64, 65]. This has also been demonstrated in rats; enteric organisms translocate on day 1 postburn to the mesenteric lymph nodes and are cleared by day 4. However, in an infected burn organisms continue to translocate to the MLN until day 7 postburn. Experimental studies in animals have produced evidence of ischemic insult to the intestine as a cause of this burn induced gut barrier failure [66]. After thermal injury vasoactive substances such as catecholamines and angiotensin in blood are elevated, causing a reduction in intestinal perfusion which is perpetuated by the hypovolemia resulting from sequestration of fluid in burned tissues. However, the splanchnic blood flow is maintained by pretreating animals with angiotensin-converting enzyme inhibitors by intra-arterial infusion of vasodilators [67]. The proposed mechanism of tissue damage after ischemic injury in these conditions is the generalization of oxygen free radicals and peroxidases [68].

Cardiopulmonary Bypass

As noted above, the gut mucosal barrier is very important in protecting the individual from the bacteria and endotoxins present in the gut lumen. Alterations in mucosal permeability allow the passage of bacteria and endotoxins into the systemic circulation, resulting in clinical endotoxemia which is associated with a high morbidity and mortality. This occurs during the hypothermic phase of cardiopulmonary bypass despite the maintenance of generally accepted levels of mean arterial blood pressure and blood flow and is probably a result of a low perfusion state and a resultant mucosal ischemia. During cardiopulmonary bypass blood shunts away from the splanchnic beds towards other vital organs which can lead to intestinal ischemia. Nonpulsatile perfusion during bypass is a stimulus for the release of angiotensin II, which is a selective splanchnic vasoconstrictor [41, 69]. It also activates platelets and leukocytes, causing the release of additional mediators that produces further vasoconstriction and endothelial damage, leading to an increase in the vascular permeability, and can contribute to the syndrome of multisystem organ failure [70]. Preexisting vascular disease enhances the development of mucosal ischemia in such situations.

Liver Diseases

Systemic endotoxemia without evidence of gram-negative infection often occurs in patients with liver disease [71]. Endotoxin concentrations in the blood also correlate with the severity of hepatic encephalopathy and risk for death [72]. Gut decontamination and endotoxin binding before liver surgery reduce the level of endotoxemia, protect the liver, and enhance postoperative survival [25].

The liver with its reticuloendothelial system plays a pivotal role in clearing toxic substances released by intestinal bacteria. Patients with end-stage liver failure often manifest high levels of circulating endotoxins, reflecting the impaired clearing capacity of the diseased liver and the effects of portosystemic shunting.

During the liver transplantation procedure endotoxemia occurs secondary to surgical manipulation, splenic hypoperfusion, portal occlusion, and the anhepatic phase. Once the liver is perfused, endotoxin levels begin to decline in parallel with lactic acid clearance [73]. However, persistence of high levels of endotoxin is related to graft nonfunction. Bacterial translocation to MLN and blood is detected in rats with portal hypertension and is increased in the ones that have suffered hemorrhagic shock [74]. Intestinal permeability is shown to be increased in patients with portal hypertension [75, 76]. All this can explain in part the higher incidence of spontaneous bacterial peritonitis encountered in these patients.

Celiac Diseases

Patients with untreated celiac disease absorb increased amounts of disaccharides and ^{51}Cr-EDTA, reflecting an increased leakiness of the mucosa at the level of intracellular junctions. They also tend to absorb less monosaccharide and PEG^{400} than do controls because of their reduced intestinal surface area. In screening patients for the presence of celiac disease the use of di/monosaccharides ratios and that of ^{51}Cr-EDTA absorption are comparable in sensitivity. Both are more sensitive than is the absorption of either d-xylose or folate. The PEG^{400} absorption method is less affected, making it a less useful test [77, 78]. Patients with dermatitis herpetiformis have changes similar to those reported in patients with celiac disease despite having a much milder form of enteropathy [79].

Treating patients with sprue or dermatitis herpetiformis with a gluten-free diet enables their intestinal permeability to normalize. PEG^{400} absorption and cellobiose/mannitol ratios normalize with treatment even before a histological recovery of the mucosa can be detected [80]. However, episodes of bacterial translocation have not been well studied in this disease.

Inflammatory Bowel Disease

Intestinal permeability tests can be used to screen patients with abdominal symptoms for inflammatory mucosal lesions. Permeability tests using lactulose/rhamnose-mannitol and PEG^{400} show increased urinary excretion ratios in the majority of cases of small-bowel, ileal, and ileocecal Crohn's disease [49, 50, 55]. The ratios return to normal in patients who respond to medical treatments. Urinary excretion of ingested ^{51}Cr-EDTA is increased in patients with small-bowel Crohn's disease, but the degree of increased urinary excretion is not closely correlated with level of disease activity, but it is so with the extent of the disease along the gastrointestinal tract. Increased intestinal permeability as assessed by ^{51}Cr-EDTA can be shown to precede clinical manifestations of Crohn's disease [81]. In patients with Crohn's disease, although they suffer increased gut permeability, the episodes of bacterial translocation which causes systemic manifestations are not significant except occasionally during periods of active disease.

Alcohol

Acute ingestion of alcohol does not affect the intestinal permeability of disaccharides; however, intoxicated patients have been shown to have increased PEG^{400} absorption, and the abnormal permeability for PEG^{400} persists for almost 2 weeks despite abstinence. Chronic alcohol ingestion also increases the absorption of ^{51}Cr-EDTA [82]. This is an important factor in patients suffering from trauma or gastrointestinal bleeding since the alcohol can perpetuate translocation and its consequences.

Current Applications of Permeability Testing

Permeability tests are simple and easy to perform. They provide accurate information regarding the integrity of the small-intestinal mucosa as a whole. They overcome the need for an intestinal biopsy in assessing the function of the small intestine in individuals with localized or patchy forms of mucosal disease. Permeability tests can be used as a diagnostic tool for ruling out small-bowel mucosal lesions and as a tool to monitor patients with mucosal diseases and their response to medical treatment. However, altered permeability is not essentially correlated with microbial translocation. The relationship between gut permeability and microbial translocation is poorly defined.

Increased gut permeability allows microbial toxins and antigens to gain access to the systemic circulation leading to different disease states. At the same time, microbial and toxin translocation through the intestinal barrier is affected by the same factors that cause altered gut permeability. However, the other physiological systems supporting the gut mucosal barrier play an important role in clearing up the micro-organisms and toxins which gain access to the MLN, liver, and/or spleen before reaching the systemic circulation. The immune system of the mucosa and the reticuloendothelial system presented in MLN, liver, and spleen as well as in the serum definitely protects patients from the complications of translocation and limits it. Multiple gastrointestinal diseases which present with increased permeability do not manifest significant translocation consequences unless the local or systemic immune system is compromised. Examples are seen in patients with celiac disease, milk intolerance, and in those ingesting nonsteroidal anti-inflammatory drugs and alcohol ingestion. However, certain pathogenic bacteria are capable of overwhelming the gut immune system, reaching the systemic circulation, and causing sepsis. Patients who are immunocompromised or malnourished can develop systemic disease due to opportunistic pathogens normally present in the gut which translocate freely and reach the blood without facing clearance by the immune system.

In conclusion, increased mucosal permeability predisposes the patient to bacterial translocation, especially when the patient's immune system is compromised. The tools used for measuring mucosal permeability can help in identifying the risks of bacterial translocation and its mechanisms. However, a combination of probes assessing different absorption pathways need to be used in the same subject to identify the specific absorptive pathways that are abnormal in each disease.

References

1. Sellin JH (1993) Intestinal electrolyte absorption and secretion. In: Sleisenger MH, Fordtran JS (eds) Gastrointestinal disease. Saunders, Pennsylvania, pp 954–971
2. Van Leeuwen PAM, Boermeester MA, Houdijk APJ et al (1994) Clinical significance of translocation. Gut 1:S28–S34
3. Dobbins WO (1982) Gut immunophysiology: a gastroenterologist view with emphasis on pathophysiology. Am J Physiol 242:91–98
4. Fink MP (1991) Gastrointestinal mucosal injury in experimental models of shock, trauma and sepsis. Crit Care Med 19:627–641
5. Hentges DJ (1983) Role of the intestinal microflora in the defense against infection. In: Hentges DJ (ed) Human intestinal microflora in health and disease. Academic, New York, pp 311–331
6. Gautreaux MD, Deitch EA, Berg RD (1990) Immunological mechanisms preventing bacterial translocation from the gastrointestinal tract. In: Heidt P (eds) Proceedings of the 10th international symposium on gnotobiology. Leiden, The Netherlands
7. Marin ML, Greenstein AJ, Geller SA, Gordon RE, Aufses AH (1983) A freeze fracture study of Crohn's disease of the terminal ileum: changes in epithelial tight junction organization. Am J Gastroenterol 78:537–547

8. Simons K, Fuller SD (1985) Cell surface polarity in epithelia. Ann Rev Cell Biol 1985;1:243–288
9. Diamond JM (1977) The epithelial junction: bridge, gate and fence. Physiologist 20:10–18
10. Madara JL, Barenberg D, Carlson S (1986) Effects of cytochalasin D on occluding junctions of intestinal absorptive cells: further evidence that the cytoskeleton may influence paracellular permeability and junctional charge selectivity. J Cell Biol 102:2125–2136
11. Claude P, Goodenough DA (1973) Fracture faces of zonulac occludentes from "tight" and "leaky" epithelia. J Cell Biol 58:390–400
12. Hollander D, Ricketts D, Boyd CAR (1988) The importance of "probe" molecular geometry in determining intestinal permeability. Can J Gastroenterol 2:35–38
13. Laker MF, Menzies IS (1977) Increase in human intestinal permeability following ingestion of hypertonic solutions. J Physiol (Lond) 273:881–894
14. Menzies IS (1984) Transmucosal passage of inert molecules in health and disease. In: Skadhange E, Heitze K (eds) Intestinal absorption and secretion. MTP, Lancaster, pp 527–543
15. Powell DW (1981) Barrier function of epithelia. Am J Physiol 241:G275–G288
16. Maxton DG, Bjarnason I, Reynolds AP, Catt D, Peters TJ, Menzies IS (1986) ^{51}Cr-labelled ethylenediaminetetra-acetate, L-rhamnose and polyethyleneglycol 500 as probe markers for assessment in vivo of human intestinal permeability. Clin Sci 71:71–80
17. Berg RD, Itoh K (1986) Bacterial translocation from the gastrointestinal tract-immunologic aspects. Microecol Ther 16:131–145
18. Walker WA (1985) Role of the mucosal barrier in toxin/microbial attachment to the gastrointestinal tract. Ciba Found Symp 112:34–47
19. Mathison JC, Ulevitch RJ (1979) The clearance, tissue distribution and cellular localization of intravenously injected lipopolysaccharide in rabbits. J Immunol 123:2133–2143
20. Maier RV, Hahnel GB (1984) Microthrombosis during endotoxemia: the role of hepatic versus alveolar macrophages. J Surg Res 36:362–370
21. Coalson JJ, Benjamin B, Archer LT et al (1978) Prolonged shock in the baboon subjected to infusion of E. coli endotoxin. Circ Shock 5:423–437
22. Van Deventer SJH, Ten Cat JW, Tytgat GNJ (1988) Intestinal endotoxemia. Gastroenterology 94:825–831
23. Schoffel U, Shiga J, Mittermayer CH (1982) The proliferation inhibiting effect of endotoxin on human endothelial cells in culture and its possible implication in states of shock. Circ Shock 9:499–508
24. Greve JW, Gouma DJ, Soeters PB, Buurman WA (1990) Suppression of cellular immunity in obstructive jaundice is caused by endotoxins. Gastroenterology 98:478–485
25. Van Leeuwen PAM, Hong RW, Rounds JD, Rodrick M, Wilmore DW (1991) Hepatic failure and coma after liver resection is reversed by manipulation of gut contents: the role of endotoxin. Surgery 110:169–175
26. Baker JW, Deitch EA, Berg R, Ma L (1987) Hemorrhagic shock impairs the mucosal barrier resulting in bacterial translocation from the gut and sepsis. Surg Forum 37:73–74
27. Deitch EA, Winterton J, Berg RD (1986) Thermal injury promotes bacterial translocation from the gastrointestinal tract in mice with impaired T cell-mediated immunity. Arch Surg 121:97–101
28. Deitch EA, Winterton J, Berg R (1987) Effect of starvation, malnutrition and trauma on the GI tract flora and bacterial translocation. Arch Surg 122:1019–1025
29. Kosiol JM, Rush BF, Smith SM et al (1988) Occurrence of bacteremia during and after hemorrhagic shock. J Trauma 28:10
30. Deitch EA, Morrison J, Berg R et al (1990) Effect of hemorrhagic shock on bacterial translocation, intestinal morphology and intestinal permeability in conventional and antibiotic decontaminated rats. Crit Care Med 18:529

31. Bailey RW, Bulkley GB, Levy KI et al (1982) Pathogenesis of nonocclusive mesenteric ischemia: studies in a porcine model induced by pericardial tamponade. Surg Forum 33:194
32. Porter JM, Sussman MS, Bulkley GB (1989) In: Marston A, BulkleyGB, Fiddian-Green BG et al (eds) Splanchnic vasospasm in circulatory shock in splanchnic ischemia and multiple organ failure. Arnold, London, pp 73–88
33. Berg RD (1992) Bacterial translocation from the gastrointestinal tract. J Med 23:217–244
34. Nolan JP (1981) Endotoxin, reticuloendothelial function and liver injury. Hepatology 1:458–465
35. Palmer HR, Duerden BI, Holdsworth CD (1980) Bacteriological and endotoxin studies in cases of ulcerative colitis submitted to surgery. Gut 21:851–854
36. Wells CL, Maddaus MA, Reynolds CM, Jechorck RP, Simmons RL (1987) Role of anaerobic flora in the translocation of aerobic and facultatively anaerobic intestinal flora. Infect Immun 55:268–269
37. Berg RD, Garlington AW (1979) Translocation of certain indigenous bacteria from the gastrointestinal tract to the mesenteric lymph nodes and other organs in a gnotobiotic mouse model. Infect Immun 23:403–411
38. Hollander D, Koyama S, Dadufalza V et al (1989) Polyethylene glycol 900 permeability of rat intestinal and colonic segments in vivo and brush border membrane vesicles in vitro. J Lab Clin Med 113:505–515
39. Menzies LS (1984) Transmucosal passage of inert molecules in health and disease. In: Skadhange E, Heintz K (eds) Intestinal absorption and secretion. MTP Press, Lancaster, pp 527–543 (Falk symposium 36)
40. Chadwick VS, Phillips SF, Hofmann AF (1977) Measurements of intestinal permeability using low molecular weight polyethylene glycols (PEG 400). Gastroenterology 73:241–246
41. Ohri SK, Bjarnason I, Pathi V et al (1993) Cardiopulmonary bypass impairs small intestinal transport and increases gut permeability. Ann Thorac Surg 55:1080–1086
42. Kamage JK, Stanisz A, Scicchitano R, Hunt RH, Perdue MH (1988) Effect of immunologic reactions on rat intestinal epithelium. Correlation of increased intestinal permeability to chromium 51-labeled ethylenediaminetetraacetic acid and ovalbumin during acute inflammation and anaphylaxis. Gastroenterology 94:1368–1375
43. Ferry DM, Butt TJ, Broom MF et al (1989) Bacterial chemotactic oligopeptides and intestinal mucosal barrier. Gastroenterology 97:61–67
44. Menzies IS (1974) Absorption of intact oligosaccharide in health and disease. Biochem Soc Trans 2:1042–1047
45. Chadwick VS, Philips SF, Hoffman AF (1977) Measurements of intestinal permeability using low molecular weight polyethylene glycols (PEG 400). II. Application to normal and abnormal permeability states in man and animals. Gastroenterology 73:247–251
46. Bjarnason I, Peters TJ, Veall N (1983) A persistent defect of intestinal permeability in coeliac disease as demonstrated by a ^{51}Cr-labelled EDTA absorption test. Lancet i:323–325
47. Ford RPK, Menzies IS, Phillips AD, Walker-Smith JA, Turner MW (1985) Intestinal sugar permeability: relationship to diarrhoeal disease and small bowel morphology. J Pediatr Gastroenterol Nutr 4:568–574
48. Philipsen EK, Batsberg W, Christensen AB (1988) Gastrointestinal permeability to polyethylene glycol: an evaluation of urinary recovery of an oral load of polyethylene glycol as a parameter of intestinal permeability in man. Eur J Clin Invest 18:139–145
49. Jenkins AP, Nukajam WS, Menzies IS, Creamer B (1992) Simultaneous administration of lactulose and ^{51}Cr-ethylenediaminetetraacetic acid: a test to distinguish colonic from small-intestinal permeability change. Scand J Gastroenterol 27:769–773
50. Katz KD, Hollander D, Vadheim CM et al (1989) Intestinal permeability in patients with Crohn's disease and their healthy relatives. Gastroenterology 97:927–931

51. Elia M, Behrens R, Northrop C, Wraight P, Neale G (1987) Evaluation of mannitol, lactulose and ^{51}Cr-labelled ethylenediaminetetraacetate as markers of intestinal permeability in man. Clin Sci 73:197–204
52. Maxton DG, Bjarnason I, Reynolds AP, Catt SD, Peters TJ, Menzies IS (1986) Lactulose, ^{51}Cr-labelled ethylenediaminetetraacetate, L-rhamnose and polyethylene glycol 400 as probe markers for assessment in vivo of human intestinal permeability. Clin Sci 71:71–80
53. Chadwick VS, Phillips SF, Hofmann AF (1977) Measurements of intestinal permeability using low molecular weight polyethylene glycols (PEG 400). Gastroenterology 73:247–251
54. Iriving CS, Lifschitz CH, Marks LM, Nichols BL, Klein PD (1986) Polyethylene glycol polymers of low molecular weight as probes of intestinal permeability, I, Innovations in analysis and quantitation. J Lab Clin Med 107:290–298
55. Jenkins RT, Ramage JK, Jones DB, Collins SM, Goodacre RL, Hunt RH (1988) Small bowel and colonic permeability to ^{51}Cr-EDTA in patients with active inflammatory bowel disease. Clin Invest Med 11:151–155
56. Morales J, Kibsey P, Thomas PD, Poznansky MJ, Hamilton S (1992) The effects of ischemia and ischemia-reperfusion on bacterial translocation, lipid peroxidation and gut histology: studies on hemorrhagic shock in pigs. J Trauma 33:221–227
57. Rush BF, Sori AJ, Murphy TF, Smith S, Flanagan JJ, Machiedo GW (1988) Endotoxaemia and bacteriaemia during hemorrhagic shock. Ann Surg 207:549–554
58. Peitzman AB, Udekwu AO, Ochoa J, Smith S (1991) Bacterial translocation in trauma patients. J Trauma 31:1083–1087
59. Parks DA, Bulkley GB, Granger DN, Hamilton SR, McCord JM (1982) Ischemic injury in the cat small intestine: role of superoxide radicals. Gastroenterology 82:9–15
60. Deitch EA, Bridges WR, Baker J et al (1988) Hemorrhagic shock-induced bacterial translocation is reduced by xanthine oxidase inhibition or inactivation. Surgery 104:191–198
61. Deitch EA, Bridges W, Ma L, Berg R, Specian R, Granger DN (1990) Hemorrhagic shock-induced bacterial translocation: the role of neutrophils and hydroxyl radicals. J Trauma 30:942–952
62. Deitch EA, Ma L, Ma J-W, Berg RD (1989a) Lethal burn-induced bacterial translocation: role of genetic resistance. J Trauma 29:1480–1487
63. Vaughan WG, Horton JW, Walker PB (1992) Allopurinol prevents intestinal permeability changes after ischemia-reperfusion injury. J Pediatr Surg 27:968–973
64. Ziegler TR, Smith RJ, O'Dwyer ST et al (1988) Increased intestinal permeability associated with infection in burn patients. Arch Surg 123:1313–1319
65. Deitch EA (1990) Intestinal permeability is increased in burn patients shortly after injury. Surgery 107:411–416
66. Morris SE, Navaratnam N, Townsend CM, Herndon DN (1988) Bacterial translocation and mesenteric blood flow in a large animal model after cutaneous thermal and smoke inhalation injury. Surg Forum 39:189–191
67. Jones WG, Minei JP, Barber AE et al (1990) Angiotensin converting enzyme inhibition decreases bacterial translocation after burn injury. FASEB J 4:A953
68. Jones WG, Barber AE, Minei JP, Fahey TJ, Shires GT, Shires GT (1991) Differential pathophysiology of bacterial translocation after thermal injury and sepsis. Ann Surg 14:24–30
69. Watkins L Jr, Lucas SK, Gardner TJ et al (1979) Angiotensin II levels during cardiopulmonary bypass: a comparison between pulsatile and nonpulsatile flow. Surg Forum 30:229–230
70. Fiddian-Green RG (1990) Gut mucosal ischemia during cardiac surgery. Semin Thorac Cardiovasc Surg 2:389–399
71. Farao K, So K, Moroi T et al (1977) Detection of endotoxin in plasma and ascites fluid of patients with cirrhosis. Its clinical significance. Gastroenterology 73:539–542
72. Bigatello LM, Broitman SA, Fattori L et al (1987) Endotoxemia, encephalopathy and mortality in cirrhotic patients. Am J Gastroenterol 82:11–15

73. DeGasperi A, DeCian W, Vaiani F et al (1994) Endotoxemia following liver transplantation in humans. Transplant Proc 26:3664–3665
74. Sorell WT, Quigley EMM, Jin G, Johnson TJ, Rikkers LF (1993) Bacterial translocation in the portal-hypertensive rat: studies in basal conditions and on exposure to hemorrhagic shock. Gastroenterology 104:1722–1726
75. Yablonski ME, Lifson N (1976) Mechanism of production of intestinal secretion by elevated venous pressure. J Clin Invest 57:904–915
76. Hamdani R, Chaparala R, Stauber R et al (1995) Intestinal permeability as measured with polyethylene glycol 600 in patients with liver cirrhosis: clinical investigation. Eur J Gastroenterol Hep (submitted)
77. Ukabam SO, Cooper BT (1984) Small intestinal permeability to mannitol, lactulose and PEG 400 in celiac disease. Dig Dis Sci 29:809–816
78. Lifschitz CH, Shulman RJ, Langston C, Gopalakrishna GS (1989) Comparison of the d-xylose and polyethylene glycol absorption tests as indicators of mucosal damage in infants with chronic diarrhea. J Pediatr Gastroenterol Nutr 8:47–50
79. Bjarnason I, Marsh MN, Price A, Levi AJ, Peters TJ (1985) Intestinal permeability in patients with coeliac disease and dermatitis herpetiformis. Gut 26:1214–1219
80. Ukabam SO, Cooper BT (1985) Small intestinal permeability as an indicator of jejunal mucosal recovery in patients with celiac sprue on a gluten-free diet. J Clin Gastroenterol 7:232–236
81. May GR, Sutherland LR, Meddings JB (1993) Is small intestinal permeability really increased in relatives of patients with Crohn's disease? Gastroenterology 104:1627–1632
82. Bjarnason I, Ward K, Peters TJ (1984) The leaky gut of alcoholism: possible route of entry for toxic compounds. Lancet 28:179–182

Serum Lactate: Marker of Prognosis, Not Inadequacy of Oxygen Delivery

J. A. Russell

Introduction

Detection of occult cellular and tissue hypoxia is fundamental to the intensivist's goal to provide optimal resuscitation to the critically ill. Plasma lactate concentration has been used as a global marker of inadequate oxygen delivery to tissues and evidence of anaerobic metabolism secondary to tissue hypoxia. Indeed, there is a relatively good direct relationship between plasma lactate levels and mortality of critical illness. However, recent studies have raised important questions regarding the specificity of lactate as a marker of tissue hypoxia. These studies include biochemical, cellular, large animal and clinical studies identifying nonhypoxic mechanisms of lactate production. There is also greater appreciation of the importance of impaired lactate clearance mechanisms in controlling plasma lactate levels in critically ill patients. In addition, several studies by our group and others have failed to show that plasma lactate predicted the dependence of global oxygen consumption (VO_2) on oxygen delivery (DO_2) when VO_2 and DO_2 were determined independently. Finally, tissue hypoxia can occur without lactic acidosis or hyperlactatemia because tissue hypoxia must be severe and sustained to cause increased lactate production and because enhanced lactate uptake by well-oxygenated cells can maintain normal plasma lactate level. Thus, the plasma lactate level is insensitive and nonspecific as a marker of tissue hypoxia. Plasma lactate may be a reasonable marker of overall prognosis but is not necessarily an accurate signal of tissue hypoxia.

Biochemistry and Physiology of Lactate Production and Clearance

Lactate Production

Pyruvate and lactate are produced by the oxidation of glucose or glycogen. The pyruvate-lactate balance is controlled by lactate dehydrogenase, $NADH/NAD^+$ ratio and oxygen availability. During anaerobic metabolism NAD^+ is not available, and reoxidation of NADH during glycolysis is therefore coupled with reduction of pyruvate to lactate [1].

During cellular hypoxia, ATP is produced as shown in Eq. 1:

$$\text{Glucose} + 2\ \text{ADP} + 2\ P_i \rightarrow 2\ \text{lactate} + 2\ \text{ATP} + 2\ H_2O \quad (1)$$

in which ADP is adenosine diphosphate and P_i is inorganic phosphate. It can be seen that this reaction does not produce protons (H^+) but only lactate. Protons (H^+) are produced by hydrolysis of ATP as shown in Eq. 2:

$$\text{ATP} + H_2O \rightarrow \text{ADP} + \text{Pi}^{+}H^{+} \quad (2)$$

Therefore, hyperlactatemia can occur without lactic acidosis. Hyperlactatemia can occur when glyclolysis is accelerated but hydrolysis of ATP is not. Accelerated glycolysis increases production of pyruvate, the immediate precursor of lactate. Increased pyruvate levels cause conversion of some pyruvate to lactate, but the lactate/pyruvate ratio remains normal, about 10:1. In contrast, during ischemia and cellular hypoxia, increased H^+ production (by hydrolysis of ATP) causes even greater conversion of pyruvate to lactate; thus the lactate/pyruvate ratio increases. As a result the lactate/pyruvate ratio is a good marker to distinguish the hyperlactatemia of accelerated glycolysis (nonhypoxic) from lactic acidosis (hypoxic). Unfortunately, pyruvate is technically difficult to measure and is rarely measured in clinical laboratories. Complete oxidative phosphorylation of glucose produces 36 mol ATP/mol glucose. In contrast, under anaerobic conditions, only 2 mol ATP is produced per mole glucose (see Eq. 1) because mitochondrial production of ATP ceases, and because the cytosol is the only source of ATP. Thus anaerobic metabolism of glucose to lactate is much less efficient than oxidative metabolism of glucose. There is a negative feedback loop during anaerobic metabolism which limits the glycolytic rate and the production of lactate. Accelerated anaerobic glycolysis during tissue hypoxia increases cellular concentrations of both lactate and hydrogen ion. Excess hydrogen ion feeds back and decreases the glycolytic rate. Thus lactate production plateaus during ischemia.

The major organs that product lactate are gut, skeletal muscle, brain, heart, lung, skin, and erythrocytes.

Stress associated with critical illness can stimulate lactate release by epinephrine-dependent mechanisms. Epinephrine stimulates the release of lactate from skeletal muscle in a dose-dependent manner for oxidation or for gluconeogenesis (Cori cycle) resulting in the hyperglycemia of stress [2, 3]. Indeed, the metabolic response to injury can be simulated by infusion of cortisol, epinephrine, and glucagon [4, 5].

Lactate Clearance

Lactate is a dead end product and must be converted back to pyruvate to be utilized for energy production (oxidized in citric acid cycle) or for gluconeogenesis. Aerobic conditions are necessary to convert lactate to pyruvate. At rest most lactate is oxidized, and only one-third is used for gluconeogenesis. Liver and kidney are the major organs of lactate clearance [6].

Descriptive Studies of Relationship of Plasma Lactate and Lactic Acidosis to Prognosis in Critically Ill Adults

It is important to distinguish hyperlactatemia from lactic acidosis. Several studies show a direct relationship between lactate concentration and mortality [7–9]. Peretz and colleagues [7] found that very high lactate concentrations were associated with 90%–100% mortality. Vincent and colleagues [9] showed that decreasing lactate levels in patients in shock are associated with a better prognosis than increasing lactate. However, the rate of decline of lactate concentrations in patients who had shock was less than the rate of decline of lactate in patients who had had a seizure. In patients who have liver disease, those in shock have higher lactate levels than those not in shock [10]; however, I believe that it is not clear how to interpret moderately increased lactate levels in patients with liver disease who have conflicting signs of shock.

We have reported as an abstract and have continued studying the relationship of changes in lactate and mortality in a prospective study of patients who have sepsis syndrome [11]. Overall 30-day mortality of these patients was 35%. About 57% of prospectively identified sepsis syndrome patients had normal lactate at onset of sepsis syndrome. About one-quarter of our patients had hyperlactatemia at baseline. Lactate fell in 80% of the patients with hyperlactatemia, and their mortality was 36%. In the 20% of patients with hyperlactatemia whose lactate levels rose, mortality was 50%. Thus there may be a graded mortality in sepsis syndrome of patients who have normal lactate, patients who have elevated lactate which decreases during resuscitation, and patients who have elevated lactate which increases during resuscitation.

Hypoxic and Nonhypoxic Mechanisms of Increased Lactate in the Critically Ill

Increased lactate is not pathognomonic of tissue hypoxia and requires a thoughtful differential diagnosis in the critically ill. I suggest that clinicians should approach hyperlactatemia as they approach hyponatremia. An abnormal laboratory result cannot be interpreted in isolation but requires review of the history, repeated clinical evaluation, and review of other laboratory data.

Major mechanisms of hyperlactatemia in the critically ill are the following:

- Hypoxic: anaerobic metabolic production of lactate
  - Global (i.e., shock)
  - Focal (e.g., bowel ischemia/infarct)
- Nonhypoxic
  - Delayed clearance of lactate
  - Arteriovenous or physiologic peripheral shunts

- Dysfunction of pyruvate dehydrogenase (PDH)
- Protein degradation causing amino acid conversion to pyruvate and then lactate
- Acclerated aerobic (as opposed to anaerobic) glycolysis

Global hypoxic mechanisms have been discussed above. It is important to emphasize that hyperlactatemia can also arise from focal sources of anaerobic production of lactate. In particular, the gut can cause hyperlactatemia and lactic acidosis because of focal cellular hypoxia of gut tissue.

Delayed Clearance of Lactate

Delayed clearance of lactate can cause hyperlactatemia without cellular hypoxia. Because hepatic and renal dysfunction are so prevalent in the critically ill, delayed clearance of lactate may be a frequent contributor to the hyperlactatemia of critical illness. I am unaware of studies using infusion of radiolabelled lactate in the critically ill to define the prevalence and severity of impaired lactate clearance during various critical illnesses such as sepsis.

Arteriovenous or Physiological Peripheral Shunts

True anatomic shunts in peripheral tissues have not been demonstrated in sepsis despite suggestive evidence including coexistence of high cardiac output and increased mixed venous oxygen saturation during sepsis. However, physiological shunts could develop because of pathological vasodilation in sepsis. Furthermore, other capillaries could have decreased perfusion and conductive flow of oxygen such that adjacent tissue is hypoxic and is generating lactate. However, the low flow state, especially during the early hypovolemic phase of sepsis, may cause sequestration of lactate in tissues, and systemic arterial plasma lactate could remain normal. During volume resuscitation increased flow could "flush" lactate from the periphery to the central circulation and increased conductive flow of oxygen could reverse tissue hypoxia. Thus a "lactate hangover" could occur during reversal of tissue hypoxia and could coexist with an increased SvO_2 (from physiological shunt) and an increased cardiac output.

Dysfunction of Pyruvate Dehydrogenase

Dysfunction of PDH can be induced in models of sepsis. PDH converts pyruvate to oxaloacetate permitting pyruvate to enter the Kreb's cycle. Dysfunction of PDH causes accumulation of pyruvate and lactate and the L/P ratio remains normal.

PDH activity is decreased in muscle of septic rats and can be partially reversed by dichloroacetate [12, 13]. Curtis and Cain [14] have shown that endotoxemic dogs develop lactic acidosis which is not secondary to tissue hypoxia but rather is caused by inactivation of PDH because dichloroacetate, a PDH activator reverses the lactic acidosis without changing DO_2 or VO_2 globally or regionally (gut and hindlimb). However, other investigators have found increased PDH activity in patients who had burns [15, 16].

Protein Degradation Causing Amino Acid Conversion to Pyruvate and Lactate

Increased protein breakdown (catabolism) in critical illness causes release of amino acids which are converted to pyruvate and lactate [17]. Hence, hyperlactatemia can occur with normal lactate/pyruvate ratio by yet another non-hypoxic mechanism.

Accelerated Aerobic (as Opposed to) Anaerobic Glycolysis

Stress increases aerobic glycolysis because of a primary abnormality of glycolysis (rather than accelerated anaerobic glycolysis). Accelerated aerobic glycolysis in excess of oxidative metabolism needs causes pyruvate to accumulate and convert to lactate and alanine (i.e., normal lactate/pyruvate ratio). Endotoxin increases mRNA synthesis and synthesis of a glucose transporter protein. The glucose transporter protein moves from its intracellular site to the plasma membrance. This glucose transporter protein increases cellular glucose uptake and release of lactate and pyruvate [18, 19].

Increased glucose transporter protein and increased release of pyruvate and lactate has been described in muscle and liver of septic rats [2].

Lactic Acidosis Without Depletion of ATP in Tissues of Septic Animals

Recent evidence suggests that sepsis causes lactic aidosis without high energy phosphate depletion indicating absence of hypoxia. Several investigators have used ^{31}P nuclear magnetic resonance spectroscopy to measure intracellular levels of P_i, ATP, and phosphocreatine (P_{Cr}). Cellular hypoxia rapidly decreases P_{Cr} and ATP and increases P_i [2]. Animal models of sepsis including endotoxemia [20] and cecal ligation and perforation [21] produced lactic acidosis without decreasing P_{Cr} or ATP in skeletal muscle.

Lactate and the Oxygen Delivery/Consumption Relationship

Dependence of global oxygen consumption (VO_2) on oxygen delivery (DO_2) has been used in clinical studies to identify occult tissue hypoxia. In studies in which VO_2 and DO_2 are both calculated [$DO_2 = CaO_2 \times CO$ and

$VO_2 = CO \times (CaO_2 - CvO_2)$] using common variables (CO and CaO_2), patients who have increased lactate have pathological dependence of VO_2 on DO_2 whereas patients who have normal lactate do not [22, 23]. In contrast, in several studies in which we determined VO_2 and DO_2 independently, increased lactate did not predict pathological dependence of VO_2 on DO_2 [24–27]. Increased lactate did not predict pathological dependence of VO_2 on DO_2 in patients who had sepsis, and in whom DO_2 was increased using dobutamine [25] or in patients who had ARDS in whom DO_2 was increased using blood transfusion [24] and using dobutamine [26].

The critical oxygen delivery was determined in humans by measuring DO_2 and VO_2 during withdrawal of inotropes and mechanical ventilation in patients in whom ICU care was discontinued. The critical DO_2 was 4.5 ml kg^{-1} min^{-1} and did not differ between patients who had increased baseline plasma lactate and those who did not. In addition, serial measurements of lactate as DO_2 decreased defined the same critical DO_2 (about 4.5 ml kg^{-1} min^{-1}) as did the VO_2/DO_2 diagram. Furthermore, many patients had increased lactate at onset of study, and while DO_2 decreased, lactate remained elevated but unchanged until the critical DO_2 was reached when lactate increased dramatically. The latter result again reinforces that critically ill patients can have hyperlactatemia without dependence of VO_2 on DO_2. Thus lactate could have been increased because of nonhypoxic causes of hyperlactatemia. We suggest that the further increase of lactate below the critical DO_2 is caused by release of lactate from hypoxic tissue. Furthermore, clearance of lactate could be impaired below the critical DO_2 which could also increase lactate below the critical DO_2 [27].

These clinical studies of the relationship of DO_2, VO_2, and lactate suggest a strategy to assess lyperlactatemia during critical ilness. If hyperlactatemia is caused by tissue hypoxia in a critically ill patient, we suggest rapid increase of DO_2 and remeasurement of lactate. A fall in lactate suggests a hypoxic cause of hyperlactatemia; however, a lack of change in lactate suggests that baseline DO_2 is adequate, and that hyperlactatemia is due to nonhypoxic causes. To date, these is no well-controlled clinical study which adequately addresses this strategy.

Clinical Implications for Monitoring Resuscitation and Judging Prognosis Using Plasma Lactate

Plasma lactate levels may be useful but should be interpreted cautiously using a differential diagnosis approach as one uses to assess patients who have hyponatremia. Hyperlactatemia and lactic acidosis are not synonymous. Arterial blood gases should be measured simultaneously with plasma lactate. In general, patients who have elevated plasma lactate levels which do not decrease during resuscitation have a poorer prognosis than patients who have normal lactate levels and patients who have high lactate levels which decrease during resuscitation.

Hyperlactatemia is neither a sensitive nor a specific marker of tissue hypoxia. Both nonhypoxic and hypoxic mechanisms of lactic acidosis should be considered in the critically ill patient. Hyperlactatemia in a septic patient may not indicate tissue hypoxia but rather could be caused by nonhypoxic mechanisms.

If a critically ill patient has hyperlactatemia or, more importantly, lactic acidosis, and the possibility of occult tissue hypoxia cannot be excluded, I perform a trial of rapidly increased DO_2 (e.g., volume, transfusion, or inotropes), and a repeat measurement of lactate should be considered. If lactate decreases, I maintain the higher DO_2. If lactate does not decrease, either alternative diagnoses should be considered (e.g., bowel ischemia/infarct, nonhypoxic mechanisms), or one should question whether the increase in DO_2 has been adequate (e.g., more than 35%). This proposed approach requires validation in a prospective clinical study.

Conclusion

In conclusion, hyperlactatemia is a nonspecific finding in critically ill patients. Hyperlactatemia could indicate hypoxic or nonhypoxic mechanisms or both. In fact, it is quite likely that many critically ill septic patients have several mechanisms of hyperlactatemia. Unfortunately, hyperlactatemia is neither specific nor sensitive for tissue hypoxia. Nonetheless, serial measurement of plasma lactate may be useful for assessing the adequacy of treatment and for determining the prognosis.

Acknowledgements. The author greatly appreciates the secretarial support of Sharon Stuart in preparation of this manuscript. I also am indebted to the faculty, fellows, and nurses in the ICUs of St. Paul's Hospital and Vancouver Hospital and Health Sciences Centre, who made all of our studies possible and rewarding. This research was supported in part by NHLBI: 1 R01HL43167-02.

References

1. Haljamae H (1987) Lactate metabolism. Intensive Care World 4:118–121
2. Hotchkiss RS, Karl IE (1992) Reevaluation of the role of cellular hypoxia and bioenergetic failure in sepsis. JAMA 267:1503–1510
3. Clutter WE, Bier DM, Shah SD, Cryer PD (1980) Epinephrine plasma metabolic clearance rates and physiologic thresholds for metabolic and hemodynamic actions in man. J Clin Invest 66:94–107
4. Bessey PQ, Watters JM, Aoki TT, Wilmore DW (1984) Combined hormonal infusion simulates the metabolic response to injury. Ann Surg 200:264–281
5. Kusaka M, Ui K (1983) Activation of the Cori cycle by epinephrine. Am J Physiol 245:E294–E302
6. Woll PJ, Record CO (1979) Lactate elimination in man: effects of lactate concentration and hepatic dysfunction. Eur J Clin Invest 9:397–404

7. Peretz DI, McGregor M, Dossetor JB et al (1964) Lactic acidosis: a clinically significant aspect of shock. Can Med Assoc J 90:673–675
8. Cady LD, Weil MH, Afifi AA, Michaeli SF, Lin VY, Shubin H (1973) Quantification of severity of critical illness with special reference to blood lactate. Crit Care Med 2:75–80
9. Vincent JL, Dufaye P, Berre J, Leeman M, Degante SP, Kahn RJ (1983) Serial lactate determinations during circulatory shock. Crit Care Med 11:449–51
10. Kruse JA, Zaidi SA, Carlson RW (1987) Significance of blood lactate levels in critically ill patients with liver disease. Am J Med 83:77–82
11. Drummond AJ, Bernard GR, Russell JA (1993) Decresing lactate is associated with lower mortality of patients who have sepsis syndrome. Chest 104:57S
12. Vary TC, Siegel JH, Nakatami T, Sato T, Aoyama H (1986) Effect of sepsis on activity of pyruvate dehydrogenase complex in skeletal muscle and liver. Am J Physiol 250:E634–640
13. Vary TC, Siegel JH, Tall BD, Morris JG (1988) Metabolic effects of partial reversal of pyruvate dehydrogenase activity by dichloroacetate in sepsis. Circ Shock 24:3–18
14. Curtis SE, Cain SM (1992) Regional and systemic oxygen delivery uptake relations and lactate flux in hyperdynamic endotoxin-treated dogs. Am Rev Respir Dis 145:348–354
15. Jahoor F, Shangraw RE, Wallfish H, Herndon DN, Wolfe RR (1989) Role of insulin and glucose oxidation in mediating the protein catabolism of burns and sepsis. Am J Physiol 257:E323–331
16. Wolfe RR, Jahoor F, Herndon DN, Miyoshi H (1991) Isotopic evaluation of the metabolism of pyruvate and related substances in normal adult volunteers and severely burned children: effect of dichloroacetate and glucose infusion. Surgery 110:54–67
17. Garber AJ, Karl IE, Kipnis DM (1976) Alanine and glutamine synthesis and release from skeletal muscle. II The precursor role of amino acids in alanine and synthesis. J Biol Chem 251:836–843
18. Zeller WP, The SM, Sweet M et al (1991) Altered glucose transporter in RNA abundance in a rate model of endotoxic shock. Biochem Biophys Res Commun 176:535–540
19. Widnell CC, Baldwin SA, Danies A, Martin S, Pasternak CA (1990) Cellular stress induces a redistribution of the glucose transporter. FASEB J 4:1634–1637
20. Jepson MM, Cox M, Bates PC et al (1987) Regional blood flow and skeletal muscle energy status in endotoxemic rats. Am J Physiol 252:E581–E587
21. Jacobs DO, Maris J, Fried R et al (1988) In vitro phosphorus 31 magnetic spectroscopy of rat hind limb, skeletal muscle during sepsis. Arch Surg 123:1425–1428
22. Fenwick JC, Dodeck PM, Ronco JJ, Phang T, Wiggs B, Russell JA (1990) Increased concentrations of plasma lactate predict pathologic dependence of VO_2 on DO_2 in patients with ARDS. J Crit Care 2:81–86
23. Vincent JL, Roman A, De Becker D, Kahn RJ (1990) Oxygen uptake/supply dependency. Am Rev Respir Dis 142:2–7
24. Ronco JJ, Phang T, Walley KR, Wiggs B, Fenwick JC, Russell JA (1991) Oxygen consumption is independent of changes in oxugen delivery in severe ARDS. Am Rev Resp Dis 143:1267–1273
25. Ronco JJ, Fenwick JC, Wiggs BR, Phang PT, Russell JA, Tweeddale MG (1993) Oxygen consumption is independent of increases in oxygen delivery by dobutamine in patients who have sepsis and normal or increased concentration of plasma lactate. Am Rev Respir Dis 147:25–31
26. Phang PT, Cunningham KF, Ronco JJ, Wiggs BR, Russell JA (1994) Mathematical coupling explains dependence of oxygen consumption on oxygen delivery in ARDS. Am J Respir Crit Care Med 150:318–323
27. Ronco JJ, Fenwick JC, Tweeddale MG et al (1993) Identification of the critical oxygen delivery for anaerobic metabolism in critically ill septic and nonseptic humans. JAMA 270:1724–1730

Augmentation of Splanchnic Blood Flow

J.-L. Vincent, D. Schmartz, and D. De Backer

Introduction

Containing around 1.4 l, or approximately 30% of the adult circulating blood flow at any given time, the splanchnic circulation represents an important blood reservoir in man [1]. In response to low cardiac output states such as those associated with hemorrhage, the celiac, superior and inferior mesenteric, and hepatic circulations are selectively reduced to maintain blood flow to the vital organs [2]. Price et al. reduced the total blood volume of conscious human volunteers by between 15% and 20% and observed preferential depletion of the splanchnic circulation; the viscera losing roughly 40% of their initial volume, compared with only 10% depletion of the central volume [1]. Activation of the renin-angiotensin system in the kidney during hypotension has been largely incriminated in this so-called autotransfusion response[2, 3]. In terms of immediate survival, selective splanchnic vasoconstriction obviously provides a natural mechanism by which severely injured organisms can maintain consciousness in the face of immediate danger. However, as artificial methods of resuscitation and life-support improve chances of survival into the long term, autotransfusion presents certain problems, since prolonged periods of splanchnic ischemia risk damage to the stomach, small and large intestines, liver, gallbladder, and pancreas [2].

Regional gut perfusion can be influenced by many substances. These include: hormones such as glucagon [4], vasoactive intestinal peptide [5], angiotensin [2, 3], and cholecystokinin [6]; digestive enzymes, toxins and, micro-organisms in the lumen; mediators such as arachidonic acid metabolites [7, 8] and platelet-activating factor (PAF) [9–11]; oxygen free radicals such as nitric oxide (NO) [12, 13] and especially superoxide agents generated by xanthine oxidase which is highly concentrated within the villus enterocyte [3, 14–16]. Some of these agents are known to have enhanced mesenteric effects. For example, the vasoconstrictive response to or angiotensin vasopressin is greater in the splanchnic than in other beds.

Importantly, the gut, and in particular the intestinal mucosa, are anatomically very susceptible to damage from hypoxia. The villi are only 20% perfused in normal conditions [15] so that the countercurrent diffusion of oxygen away from the villus tip may be particularly deleterious during acute circulatory failure. Injury at the villus tip then spreads to the deeper mucosal layers with increasing degree or duration of insult. One of the first effects of this damage is increased permeability to large molecules. Damage to the mucosal layer, subepithelial damage, shedding of villus tip cells, and eventually mucosal necrosis follow [2].

Sepsis and organ failure have been linked to decreased mesenteric perfusion. In zymosan-treated rats Scalia et al. observed a possible connection between the inflammatory response and mesenteric vasoconstriction [17]. Moreover, it is well known that ischemia renders the gut more permeable to bacteria and the products of translocation. These in turn may stimulate the immune system and hence the release of cytokines and other mediators. This vicious cycle of ischemia, translocation, inflammation, and hypotension has led some experts to identify the gut as the "motor" of multiple-organ failure [18].

Inflammatory mediators have been identified on numerous occasions as playing a central role in the development of distant organ failure (i.e., far from the original source of infection/translocation). For example, the circulatory link between the liver and the intestine provides a pathway for hepatic damage following mesenteric ischemia. Furthermore, hypoxia markedly reduces the effectiveness of the liver in defending against intestinally released factors [19]. Translocation is probably less common in humans than in animals, but it has nevertheless been well documented in patients and especially those with intestinal inflammatory disease. However, translocation-induced inflammation is not a prerequisite for considering the gut as a motor of multiple-organ failure if the gut itself is an important source of mediators.

Gut mucosal damage results from several mechanisms, many of which are associated with the septic response. Such deleterious mechanisms are mediated through: cytokines, oxygen species, neutrophil derived enzymes, nutritional substrates, lipids, and intracellular hypoxia. Of these, the latter one is thought to be very important to increased gut permeability.

Damage to the splanchnic region is not confined to septic states. Reperfusion after hypoxia has also been identified as a cause of splanchnic injury [16, 20, 21]. Oxygen free radicals have a key part to play in this pathophysiology.

Splanchnic Blood Flow in Sepsis

In fact, studies vary widely with respect to their observations of blood flow distribution during acute conditions. Species, method of anesthesia and degree of fluid loading are just a few of the many influential factors. The only congruent finding is that myocardial blood flow remains consistent [22, 23]. In some studies blood flow to splanchnic organs was not decreased, indicative of marked splanchnic vasodilation in sepsis [22]. This is contrast to typical findings of increased splanchnic vascular resistance seen during experimental hemorrhagic and cardiogenic shock. Nevertheless, that mesenteric hypoperfusion may persist despite global hemodynamic improvement following resuscitation has been well documented. For example, Meadow and Meus observed this phenomenon in a piglet model of streptococcal sepsis [24], while in humans Maynard et al. suspected a similar etiology based on tonometric measurements of gastric intramucosal pH [25].

In animal studies the effects of acute sepsis on the distribution of blood flow can result in a selective impairment of the splanchnic blood supply. Indeed, in studying a mesenteric blood flow decrease associated with preserved cardiac index, Fink et al. noted that in the pig, a lipopolysaccharide-induced decrement in oxygen delivery (DO_2) was sufficient to cause ischemia but that a mechanical reduction in blood flow did not have the same effect [26]. This led them to hypothesize that endotoxemia may increase the metabolic demands of the gut epithelium, thus diminishing its endurance to decreases in DO_2 that ordinarily would be tolerated. This hypothesis is supported by the findings of Arvidsson et al. who used a porcine model to compare the effects of septic and hemorrhagic shock models on total, gastrointestinal, and liver DO_2 and oxygen uptake (VO_2) [19]. They demonstrated that gastrointestinal and hepatic VO_2 were increased significantly in septic shock but not in hemorrhage. Vallet et al. found that endotoxemia in dogs impaired the ability of individual organ tissues to optimize O_2 extraction in response to reduced DO_2 via interfering with the distribution of blood flow according to demand [27]. By inferring blood flow from measurements of oxygen pressure these authors found that the gut is less able to regulate blood flow than the whole body in general in response to lipopolysaccharide infusion. Also in dogs, Nelson et al. measured local O_2 extraction in the whole body and in an autoperfused loop of small intestine [28]. Endotoxin administration produced a O_2 supply dependency relation that occurred at a significantly higher DO_2, indicating altered oxygen extraction capabilities in this region.

Similarly, in hemodynamically stable septic patients Ruokonen et al. found sepsis-related increases in both splanchnic blood flow and splanchnic VO_2, but the latter was too extreme to be compensated by the former [29].

In summary, although the gut can increase its oxygen extraction by up to sixfold [30], thereby maintaining relatively normal levels across a broad range of flows, supply dependency inevitably results below a critical flow level. In sepsis the extraction capabilities of the gut are further impaired, and splanchnic ischemia results at a higher critical level. Concurrently, increased metabolism renders the gut more susceptible to a reduction in blood flow. The effects of sepsis on oxygen delivery to the gut are quite variable, but a reduction in splanchnic blood flow can be overwhelmed by sepsis-induced maldistributive defects, so that mesenteric hypoperfusion may persist despite apparent hemodynamic stability. Maintenance of splanchnic blood flow is essential to prevent the interplay between inadequate oxygen supply and inflammatory reaction that could cause an increase in gut permeability and the release of the mediators of sepsis.

How To Augment Splanchnic Blood Flow?

Fluid Resuscitation

Commencement of fluid administration is the essential first step in resuscitative therapy. Ottoson compared the effects of treating septicemic rats with antibiotics only, corticosteroids only, and Ringer's lactate only versus combinations of these agents [31]. In the group that received antibiotics alone, survival was 10%; in the other single medication groups survival was between 50% and 70%; combining these agents, however, increased survival to 90%.

Administration of a colloid, such as hetastarch, is an important factor in maintaining splanchnic blood flow. In a porcine model Fink et al. reported that the most successful resuscitating technique required combination of colloid plus Ringer's lactate plus dobutamine [26]. Sufficient volemia is thus very important but may not be sufficient; adrenergic agents also have a part to play.

Adrenergic Agents

In general, β-adrenergic stimulation increases splanchnic blood flow while α stimulation is responsible for a decrease in splanchnic blood flow (Fig. 1). Dobutamine is a β stimulator with only mild effects on peripheral vasculature. Norepinephrine has a mainly peripheral vasoconstricting effect and may cause an increase in systemic vascular resistance and a decrease in peripheral vascular capillaries. Ross examined the effects of epinephrine and norepinephrine on the feline mesenteric circulation [32]. He was able to produce mesenteric vasodilation, at least in part, by stimulating β-adrenergic receptors. In anesthetized dogs Charbon and Reneman studied the effects of isoproterenol [33], which has a very potent vasodilating interaction with β-receptors. Mean arterial pressure and heart rate were altered only slightly,

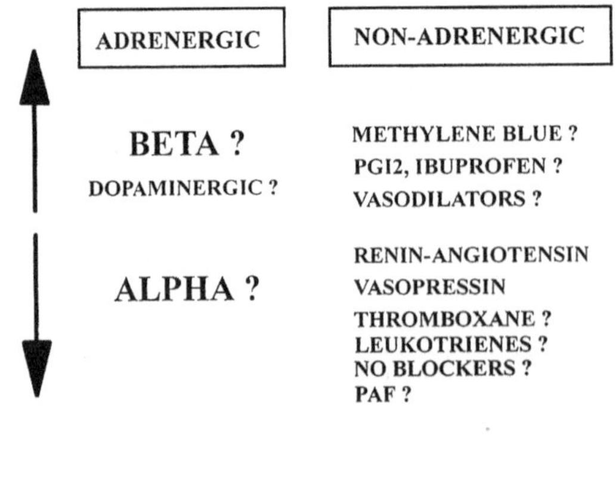

Fig. 1. Adrenergic and nonadrenergic options in response to splanchnic ischemia

except at high doses. The superior mesenteric artery exhibited a considerable dose-dependent increase in flow. Importantly, the effects could be blocked by propranolol and reduced by alprenolol, incriminating β stimulation as opposed to chemical effects in this study.

Several studies, however, infer that the effects of most adrenergic agents on the regional blood flow in sepsis are quite similar. For example, Schneider et al. found dobutamine and norepinephrine to be equivalent in a porcine model of endotoxin-induced septic shock [34]. They made the general observation that large amounts of fluid must be infused to restore ventricular preload and cardiac output, ultimately leading to a hyperdynamic state.

Dopaminergic agents are intended to increase splanchnic and renal blood flow selectively. This has been shown in many pharmacological studies, but the effects are less impressive in sepsis. Bersten et al. studied dopamine, dobutamine, and norepinephrine administration in septic and nonseptic awake sheep [23]. In general there were no major differences between any of the adrenergic agents. Breslow et al. observed similar effects whether using dopamine or other vasoconstrictors such as norepinephrine or phenylephrine in a porcine model of fluid-resuscitated septic shock [22].

Segal et al. used a model of hemorrhagic shock in anesthetized pigs to study the effects of dopamine [35]. The rational was to attempt to delay the onset of gut ischemia in septic shock by a low-dose ($2\ \mu g\ kg^{-1}\ min^{-1}$) administration. In fact, low-dose dopamine decreased the oxygen extraction capabilities of gut. Differences in animal species and surgical preparation (e.g., whole gut in situ vs. exteriorized segment) were put forward as possible explanations of the difference between this and other studies. In the unanesthetized lamb Feltes et al. found that dopamine did not improve distribution of blood flow in lambs [36]. Indeed, they found that high doses may even lower blood flow to the gut. They concluded that these differences may be related to differences in adrenergic receptors of the very young animal, and namely that the α-receptors are developed just before birth while the β_2-receptors require postnatal maturation.

Ruokonen et al. studied the effect of vasopressor doses of dopamine and norepinephrine on hemodynamics, blood flow distribution, and oxygen transport during the treatment of ten patients with hyperdynamic septic shock. For comparison, postoperation cardiac patients were selected as a control to represent ventilation changes. Splanchnic and leg blood flows were measured using indocyanine green. Either dopamine or norepinephrine was administered in random order to achieve a mean arterial pressure higher than 70 mmHg. Splanchnic flow increased in all dopamine-treated patients with septic shock, but in the norepinephrine-treated patients the results were nonuniform. Dopamine increased splanchnic DO_2, but the effect on splanchnic VO_2 was not statistically significant. In most cases the regional changes in blood flow and VO_2 were parallel, suggesting flow dependence of VO_2 associated with impaired extraction. The authors found that this was indicative of increased demand in excess of enhanced flow, and that dopamine and norepinephrine exhibited similar systemic hemodynamic effects.

With its predominant dopaminergic an β-adrenergic effects, dopexamine stimulation is expected to increase blood flow in the splanchnic vascular bed, while due to the absence of α effects, it should not cause vasoconstriction. In nonendotoxemic dogs Bredle and Cain [37] observed that dopexamine administration was associated with an increase in the critical DO_2 and a decrease in the critical oxygen ratio in resting skeletal muscle comparable with the values seen in endotoxin-treated animals. This was attributed to a loss of balance between the vasoconstrictor and vasodilator forces that normally act to direct blood flow to areas of greater oxygen demand. In endotoxin plus dopexamine dogs the critical oxygen extraction and DO_2 were not significantly different from the endotoxin alone group. Finally, systemic VO_2 was better maintained in the endotoxin plus dopexamine group than in the group that received endotoxin alone.

Van Lambalgen et al. compared dopexamine and dobutamine treatment in endotoxin-treated rats. No major differences were found between the two agents, and acute renal failure and shock developed whether the adrenergic stimulation was commenced before [38] or simultaneously [39] with endotoxin.

Tighe et al., however, found that in anesthetized pigs dopexamine was perhaps responsible for an augmented portal blood flow, and that there was preservation of relatively normal hepatic structure and maintenance organ perfusion [40]. In a similar model Webb et al. found that hemodynamic function was not significantly different between the groups, but that there was a suggestion of improvement in hepatic function [41]. Interestingly, in a study of anesthetized dogs, Cain and Curtis concluded that dopexamine temporarily increased systemic DO_2 and that VO_2 may have been responsible for preferential perfusion of the intestinal mucosa, as indicated by a nil production of lactate in the gut [42].

In summary, adrenergic agents have very little effect on the distribution of blood flow especially in septic conditions. Adrenergic agents acting predominantly to increase blood flow such as dobutamine are most effective to increase splanchnic blood flow.

Renin-Angiotensin

As has been discussed, angiotensin has been widely implicated in reduced splanchnic blood flow. It is possible to block the effects of angiotensin with an angiotensin-converting enzyme inhibitor (ACE). Administration of such an inhibitor may be beneficial to gut blood flow, but in septic shock this may be hazardous since its vasodilating effects would lead to decreased arterial pressure.

Arachidonic Acid Metabolites

The metabolites of arachidonic acid (prostaglandins, thromboxane A_2, leukotrienes) may have both beneficial and deleterious effects, and related therapies are therefore intended to inhibit one or both of the pathways or to enhance the beneficial effects and/or reduce the deleterious effects.

Cyclo-oxygenase inhibitors can improve long-term survival in some models. In a porcine model of aggressively resuscitated endotoxic shock ibuprofen and meclofenamate increased systemic and mesenteric DO_2 and diminished systemic VO_2 [43]. Splanchnic metabolism probably remained elevated, however, as indicated by the presence of mucosal acidosis. These findings were confirmed in a later study in which ibuprofen improved short-term survival and transiently ameliorated lipopolysaccharide-induced mesenteric hypoperfusion and ileal mucosal acidosis [44].

Nishijma et al. studied the effects of ibuprofen and naloxone in an anesthetized porcine model of endotoxic shock [45]. Endotoxin decreased arterial pressure and also blood flow to the kidney, spleen, and other areas. Arterial pressure was increased by both ibuprofen and naloxone but to a greater extent by the latter. Ibuprofen increased renal blood flow, and this was attributed either to increased renal perfusion or decreased production of vasoconstricting prostaglandins. Naloxone had no effect on regional blood flow.

The potential beneficial effects of prostaglandin therapy were explored by Lefer et al. in anesthetized cats [46]. Following *Escherichia coli* administration mean arterial pressure and superior mesenteric artery blood flow decreased markedly. Prostacyclin infusion dilated the splanchnic circulation and significantly increased the superior mesenteric blood flow. Lefer et al. discussed the ability of prostaglandins in general to prevent some of the pathological alterations associated with endotoxemia. However, they added that some of the effects of prostaglandins, such as increased heart rate and splanchnic constriction may not be beneficial in very acute conditions. In particular, prostacyclin, or its congener prostaglandin E_1, can increase blood flow to the splanchnic region and increase cardiac output.

Leukotrienes, on the other hand, can decrease splanchnic blood flow and impair gut perfusion. In an anesthetized porcine model of endotoxemia Cohn et al. studied the effects of LY171883, a leukotriene inhibitor [47] and noted that while the drug had no appreciable effect in normal animals, it could substantially improve mesenteric blood flow and DO_2 in endotoxic animals.

Antioxidants

Oxygen free radicals have been extensively implicated in reperfusion injury following splanchnic ischemia [45]. Zhang et al. studied the effects of pretreatment with *N*-acetyl-l-cysteine in anesthetized endotoxemic dogs [48]. The oxygen extraction capabilities of the pretreated animals were greater than those which received endotoxin alone. Peak lactate levels and tumor necrosis factor-α levels were also lower in these animals. There was, however, no beneficial effect on mesenteric flow [48]. In another study of anesthetized dogs Zhang and coworkers recently reported that the lazaroid U74006F can improve oxygen extraction capabilities in endotoxic shock. They also observed that this agent does indeed increase mesenteric and renal blood flow following endotoxin administration [49].

Effects of NO-Blocking Agents

NO blockade requires further study. It has been proposed that administration of NO blocking agents could help to increase vascular tone and mesenteric perfusion pressure, therefore improving mesenteric perfusion [50]. However, some studies suggest that NO induced synthase inhibitor, N^{G}-monomethyl-l-arginine, decreases mesenteric perfusion [51, 52], and that NO blockade can increase mesenteric sensitivity to circulating and locally produced vasoconstrictors [12]. Nevertheless, Zhang et al. found that inhibiting soluble guanylate cyclase activity with methylene blue can selectively increase mesenteric blood flow [53].

PAF Receptor Antagonists

Since PAF has been linked to gut mucosal damage in sepsis [54, 55], PAF receptor antagonists may be able to ameliorate alterations. However, results have been equivocal to date, and further study is required in this area.

Nutrition

Oral nutrition must be considered carefully as a therapeutic option. While it can increase gut DO_2, VO_2 is also increased so that the overall effect may be deleterious. For this reason oral feeding should probably be avoided in acute circulatory failure but resumed as soon as possible after resuscitation.

Conclusion

Autotransfusion of splanchnic blood to the vital organs represents an effective, albeit primitive, method of short-term survival following severe hemorrhage (Fig. 2). However, in acute conditions, and in particular during acute circulatory failure and immunological stimulation, this same mechanism may, on balance, do more harm than good. Once any immediate threat has been surpassed, the subsequent risk of splanchnic necrosis and further immunological stimulation, whether through translocation or other pathways carries a very high risk of distant multiple-organ failure and septic shock, for which the prognosis is not encouraging.

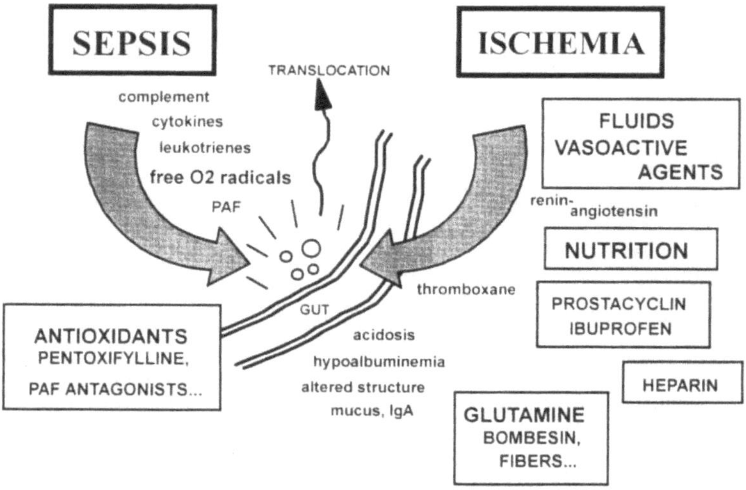

Fig. 2. Interrelation of splanchnic ischemia, translocation, and sepsis, with possible therapeutic options

Augmenting and maintaining splanchnic blood flow is therefore a fundamental prerequisite of effective therapy for the acutely ill patient and should be considered as a key goal of any resuscitative practice. To this end, fluid treatment must aim to maximize both systemic and regional perfusion. Adrenergic agents are necessary to improve vascular tone and blood distribution, but doing this selectively with, for example, dopamine and dopexamine, has not proven efficacious. At present the best approach may be to stimulate global vasodilation with dobutamine. Therapies targeted at the arachidonic metabolites, oxygen free radicals, and other mediators of the immunological response including nitric oxide may eventually join the list of effective treatments, but their use needs further study. Finally, oral nutrition must not be discounted, depending on the extent of sepsis.

References

1. Price HL, Deutsch S, Marshall BE, Stephen GW, Behar MG, Neufeld GR (1966) Hemodynamic and metabolic effects of hemorrhage in man, with particular reference to the splanchnic circulation. Circ Res 18:469–474
2. Reilly PM, Bulkley GB (1993) Vasoactive mediators and splanchnic perfusion. Crit Care Med 21:S55–S68
3. Xu D, Qi L, Guillory D, Cruz N, Berg R, Deitch EA (1993) Mechanisms of endotoxin-induced intestinal injury in a hyperdynamic model of sepsis. J Trauma 34:676–682
4. Warrwick M, Lin TM (1975) Action of glucagon and atropine on mucosal blood flow of resting fundic pouches of dogs. Life Sci 17:333–338
5. Mailman D (1978) Effects of vasoactive intestinal polypeptide on intestinal absorption and blood flow. J Physiol (Lond) 279:121–132

6. Schuurkes JA, Charbon GA (1978) Motility and hemodynamics of the canine gastrointestinal tract. Stimulation by pentagastrin, cholecystokinin and vasopressin. Arch Int Pharmocodyn Ther 236:214–227
7. Fink MP, Cohn SM, Lee PC, Rothschild HR, Deniz YF, Wang H, Fiddian-Green RG (1989) Effect of lipopolysaccharide on intestinal intramucosal hydrogen ion concentration in pigs: evidence of gut ischemia in a normodynamic model of septic shock. Crit Care Med 17:641-646
8. Fink MP (1993) Adequacy of gut oxygenation in endotoxemia and sepsis. Crit Care Med 21:S4–S8
9. Hsueh W, Gonzalez-Crussi F, Arroyave JL (1987) Platelet-activating factor: an endogenous mediator for bowel necrosis in endotoxemia. FASEB J 1:403–411
10. Sun XM, Hsueh W (1988) Bowel necrosis induced by tumor necrosis factor in rats is mediated by platelet-activating factor. J Clin Invest 81:1328–1332
11. Gonzalez-Crussi F, Hsueh W (1983) Experimental model of ischemic bowel necrosis. The role of platelet-activating factor and endotoxin. Am J Pathol 112:127–131
12. Mitchell JA, Kohlhaas KL, Sorrentino R, Warner TD, Murad F, Vane JR (1993) Induction by endotoxin of nitric oxide synthase in the rat mesentery: lack of effect on action of vasoconstrictors. Br J Pharmacol 109:265–270
13. Szabo C, Mitchell JA, Thiemermann C, Vane JR (1993) Nitric oxide-mediated hyporeactivity to noradrenaline precedes the induction of nitric oxide synthase in endotoxin shock. Br J Pharmacol 108:786–792
14. Parks DA, Granger DN (1986) Xanthine oxidase: biochemistry distribution and physiology. Acta Physiol Scand 548:97–100
15. Granger DN, Kvietys PR, Korthuis RJ, Premen AJ (1989) Microcirculation of the intestinal mucosa. In: Rauner BB, Wood JD, Schultz SG (eds) Handbook of physiology; the gastrointestinal system. Oxford University Press, New York, pp 1405–1474
16. Granger DN, Rutili G, McCord JM (1981) Superoxide radicals in feline intestinal ischemia. Gastroenterology 81:22–29
17. Scalia S, Sharma P, Rodriguez J, Roche F, Luchette F, Chambers R, Flint LM, Steinberg S (1992) Decreased mesenteric blood flow in experimental multiple organ failure. J Surg Res 52:1–5
18. Carrico CJ, Meakins JL, Marshall JC, Iry D, Maier RV (1986) Multiple-organ-failure syndrome. Arch Surg 121:196–208
19. Arvidsson D, Rasmussen I, Almqvist P, Niklasson F, Haglund U (1991) Splanchnic oxygen consumption in septic and hemorrhagic shock. Surgery 109:190–197
20. Parks DA, Granger DN (1986) Contributions of ischemia and reperfusion to mucosal lesion formation. Am J Physiol 250:G749–G755
21. Parks DA, Granger DN (1983) Ischemia-induced vascular changes: role of xanthine oxidase and hydroxyl radicals. Am J Physiol 245:G285–G291
22. Breslow MJ, Miller CF, Parker SD, Walman AT, Traystman RJ (1987) Effect of vasopressors on organ blood flow during endotoxin shock in pigs. Am J Physiol 252:H291–H300
23. Bersten AD, Hersch M, Cheung H, Rutledge FS, Sibbald WJ (1992) The effect of various sympathomimetics on the regional circulations in hyperdynamic sepsis. Surgery 112:549–561
24. Meadow WL, Meus PJ (1985) Unsuspected mesenteric hypoperfusion despite apparent hemodynamic recovery in the early phase of septic shock in piglets. Circ Shock 15:123–129
25. Burlew BS, Gheorghiade M, Jafri SM, Goldberg AD, Goldstein S (1987) Acute and chronic hemodynamic effects of nicardipine hydrochloride in patients with heart failure. Am Heart J 114:793–804
26. Fink MP, Kaups KL, Wang H, Rothschild HR (1991) Maintenance of superior mesenteric arterial perfusion prevents increased intestinal mucosal permeability in endotoxic pigs. Surgery 110:154–161
27. Vallet B, Lund N, Curtis SE, Kelly D, Cain SM (1994) Gut and muscle tissue PO2 in endotoxemic dogs during shock and resuscitation. J Appl Physiol 76:793–800

28. Nelson DP, Beyer C, Samsel RW, Wood LDH, Schumacker PT (1987) Pathological supply dependence of O2 uptake during bacteremia in dogs. J Appl Physiol 63:1487–1492
29. Ruokonen E, Takala J, Kari A, Saxen H, Mertsola J, Hansen EJ (1993) Regional blood flow and oxygen transport in septic shock. Crit Care Med 21:1296–1303
30. Kvietys PR, Granger DN (1982) Relationship between intestinal blood flow and oxygen uptake. Am J Physiol 242:G202–G209
31. Ottosson J, Dawidson I, Brandberg A, Idvall J, Sandor Z (1991) Cardiac output and organ blood flow in experimental septic shock: the effect of treatment with antibiotics, corticosteroids, and fluid infusion. Circ Shock 35:14–24
32. Ross G (1967) Effects of epinephrine and norepinephrine on the mesenteric circulation of the cat. Am J Physiol 212:1037–1042
33. Charbon GA, Reneman RS (1970) The effects of β-receptor agonists and antagonists on regional blood flow. Eur J Pharmacol 9:21–26
34. Schneider AJ, Groeneveld AB, Teule GJ, Wesdorp RI, Thijs LG (1991) Total body blood volume redistribution in porcine E. coli septic shock: effect of volume loading, dobutamine and norepinephrine. Circ Shock 35:215–222
35. Segal JM, Phang T, Walley KR (1992) Low-dose dopamine hastens onset of gut ischemia in a porcine model of hemorrhagic shock. J Appl Physiol 73:1159–1164
36. Feltes TF, Hansen TN, Martin CG, Leblanc AL, Smith S, Giesler ME (1987) The effects of dopamine infusion on regional blood flow in newborn lambs. Pediatr Res 21:131–136
37. Bredle DL, Cain SM (1991) Systemic and muscle oxygen uptake/delivery after dopexamine infusion in endotoxic dogs. Crit Care Med 19:198–202
38. Van Lambalgen AA, Van Kraats AA, Van de Vaart-Mulder MF, Van den Bos GC, Stel HV, Straub JP, Doncker AJM, Thijs LG (1991) Systemic and renal actions of dopexamine and dobutamine during endotoxin shock in the rat. J Crit Care 6:61–70
39. Van Lambalgen AA, Van Kraats AA, Mulder MF, Van den Bos GC, Teerlink T, Thijs LG (1993) Organ blood flow and distribution of cardiac output in dopexamine- or dobutamine-treated endotoxemic rats. J Crit Care 8:117–127
40. Tighe D, Moss R, Haywood G, Webb A, Al-Saady N, Heath F, Bennett D (1993) Dopexamine hydrochloride maintains portal blood flow and attenuates hepatic ultrastructural changes in a porcine peritonitis model of multiple system organ failure. Circ Shock 39:199–206
41. Webb AR, Moss RF, Tighe D, Al-Saady N, Bennett ED (1992) The effects of dobutamine, dopexamine and fluid on hepatic histological responses to porcine faecal peritonitis. Intensive Care Med 17:487–493
42. Cain SM, Curtis SE (1991) Systemic and regional oxygen uptake and delivery and lactate flux in endotoxic dogs infused with dopexamine. Crit Care Med 19:1552–1560
43. Fink MP, Rothschild R, Deniz YF, Wang H, Lee PC, Cohn SM (1989) Systemic and mesenteric O2 metabolism in endotoxic pigs: effect of ibuprofen and meclofenamate. J Appl Physiol 67:1950–1957
44. Fink MP, Kaups KL, Wang H, Rothschild HR (1992) Ibuprofen improves survival but does not ameliorate increased gut mucosal permeability in endotoxic pigs. Arch Surg 127:49–54
45. Nishijima MK, Breslow MJ, Miller CF, Traystman RJ (1988) Effect of naloxone and ibuprofen on organ blood flow during endotoxic shock in pig. Am J Physiol 255:H177–H184
46. Lefer AM, Tabas J, Smith EF (1980) Salutary effects of prostacyclin in endotoxic shock. Pharmacology 21:206–212
47. Etemadi AR, Tempel GE, Farah BA, Wise WC, Halushka PV, Cook JA (1987) Beneficial effects of a leukotriene antagonist on endotoxin-induced acute hemodynamic alterations. Circ Shock 22:55–63
48. Zhang H, Spapen H, Nguyen DN, Rogiers P, Bakker J, Vincent JL (1995) Effects of N-acetyl-L-cysteine on regional blood flow during endotoxic shock. Eur J Surg 9:236–243

49. Zhang H, Spapen H, Manikis P, Rogiers P, Metz G, Buurman WA, Vincent JL (1995) Tirilazad mesylate (U74006F) inhibits the effets of endotoxin in dogs. Am J Physiol 268:H1847–H1855
50. Spain DA, Wilson MA, Bar-Natan MF, Garrison RN (1994) Nitric oxide synthase inhibition aggravates intestinal microvascular vasoconstriction and hypoperfusion of bacteremia. J Trauma 36:720–725
51. Meyer J, Traber LD, Nelson S, Lentz CW, Nakazawa H, Herndon DN, Noda H, Traber DL (1992) Reversal of hyperdynamic response to continuous endotoxin administration by inhibition of NO synthesis. J Appl Physiol 73:324–328
52. Hutcheson IR, Whittle BJR, Boughton-Smith NK (1990) Role of nitric oxide in maintaining vascular integrity in endotoxin-induced acute intestinal damage in the rat. Br J Pharmacol 101:815–820
53. Zhang H, Rogiers P, Preiser JC, Spapen H, Manikis P, Vincent JL (1995) Effects of methylene blue on oxygen availability and regional blood flow during endotoxic shock. Crit Care Med (in press)
54. Chang SW, Feddersen CO, Henson PM, Voelkel NF (1987) Platelet-activating factor mediates hemodynamic changes and lung injury in endotoxin-treated rats. J Clin Invest 79:1498–1509
55. Sun X, Hsueh W (1988) Bowel necrosis induced by tumour necrosis factor in rats is mediated by platelet activating factor. J Clin Invest 81:1328–1331

Selective Decontamination of the Digestive Tract

C. P. Stoutenbeek and H. K. F van Saene

Introduction and Terminology

The conventional classification of hospital infections is into *community-acquired* and *nosocomial* infections, the latter referring to infections developing after 48 h following admission to the hospital. A further period of 48 h is commonly used to distinguish between intensive care unit (ICU)-acquired infections and infections incubating after admission to the ICU (primary infections). Primary infections may thus be either nosocomial or community acquired, for example, in trauma patients admitted directly to the ICU without prior hospital admission.

In terms of preventing infection it is more useful to distinguish between *exogenous* and *endogenous* infections. Conventional infection prevention policies are mainly directed against exogenous infections, whereas selective decontamination of the digestive tract (SDD) is a method to prevent endogenous infections.

Endogenous infections are defined as those caused by potentially pathogenic micro-organisms (PPM) and carried in the throat and gastrointestinal (GI) tract, whereas exogenous infections are caused by PPM that have multiplied outside the body (e.g., a contaminated nebulizer). Endogenous infections can be further subdivided into *primary* and *secondary* endogenous infections. Primary endogenous infections are caused by PPM carried in the throat or GI tract on admission to the ICU, whereas secondary endogenous infections are caused by PPM that have colonized the throat and GI tract during the ICU stay. Although both exogenous infections and secondary endogenous infections are caused by hospital-acquired micro-organisms, the difference is that in exogenous infections the multiplication of PPM, which is an essential step in the pathogenesis of infections, takes place outside the body and in secondary endogenous infections inside the oropharynx or GI tract. This difference is important from the point of view of prophylaxis because the multiplication of pathogens inside the body can be prevented by SDD.

Basic Principles

SDD is a technique for preventing infections in granulocytopenic patients and critically ill intensive care patients [5]. It is based on two observations:

firstly, that most infections are either primary or secondary endogenous and, secondly, that the indigenous, mostly anaerobic flora have important physiological functions and should be preserved as much as possible. The technique is called selective because the topical antibiotics used in SDD have little effect on the indigenous flora. Aerobic gram-positive cocci (enterococci and coagulase negative staphylococci) belong to the indigenous flora. The pathogenic potential is somewhat higher than of anaerobes but much lower than that of *S. aureus*, aerobic gram-negative PPM, or yeasts.

SDD includes three essential elements: (a) the application of topical nonabsorbable antibiotics to both the oropharyngeal cavity and GI tract during ICU stay, (b) a short period of systemic antibiotic prophylaxis in the first few days, and (c) intensive bacteriological monitoring of the oropharynx and GI tract.

Topical Nonabsorbable Antibiotics

The goal of using topical antibiotics is to selectively eliminate aerobic PPM from the oropharyngeal cavity and GI tract, thereby preventing endogenous infections and emergence of resistance. The topical antibiotics used for SDD include polymyxin, tobramycin, and amphotericin B (PTA). The PTA regimen covers all aerobic gram-negative PPM, *S. aureus*, and yeasts but has relatively little effect on the indigenous flora (i.e. *selective* decontamination). Polymyxin is the essential component because of its broad gram-negative spectrum and because acquired resistance to polymyxin is virtually nonexistent. However, the gap in the spectrum of polymyxin includes *Proteus* and *Providencia* spp., and therefore it should be combined with tobramycin. Tobramycin has the additional advantage of potentiating the effect of polymyxin on *Pseudomonas* and *Acinetobacter* spp. The obvious gap in the spectrum with the combination of polymyxin and tobramycin is methicillin-resistant-*Staphylococcus aureus* (MRSA), which is often also resistant to aminoglycosides. The correct choice and dosage of topical antibiotics is crucial for success of the decontamination because many antibiotics are less suitable for decontamination as they are inactivated by feces or by concomitant medication such as antacids and sucralfate [6].

The oropharynx is the most important source of PPM causing respiratory tract infections and infections in the upper part of the body (e.g. line infections), whereas the GI tract is more important for urinary tract infections, abdominal infections and wounds in the lower part of the body. It therefore is of the utmost importance to decontaminate both the oropharynx and the GI tract.

In an effectively decontaminated patient only anaerobes, enterococci, viridans streptococci and coagulase negative staphylococci should be found in throat or rectal culture. It should be realized that this is not the consequence of overgrowth by gram positive cocci but the aim of selective elimination of

PPM. The isolation of enterococci and coagulase negative staphylococci in the oropharynx is a normal consequence of this method without clinical sequelae.

Systemic Antibiotic Prophylaxis

Primary endogenous infections, occurring within the first 4 days of ICU stay constitute about 50% of all respiratory infections occurring in ICUs when no antibiotic prophylaxis is given [7].

Infection prevention with SDD should be combined with short-term systemic antibiotic prophylaxis in the first 3–4 days of ICU stay in order to (a) prevent primary endogenous infections in the first few days when the decontamination is not yet complete, and (b) eliminate gram-positive PPM from the oropharynx (e.g., pneumococci, hemolytic streptococci) which are insensitive to the topical antibiotics.

Microbiological Surveillance

Bacteriological monitoring is essential for the success of SDD: (a) to control the effectiveness of the decontamination procedure, (b) to adjust the antibiotic regimen in case of colonization by resistant strains (e.g., MRSA), and (c) to recognize and prevent exogenous infections.

Systematic Reviews

To date there have been more than 40 trials on SDD. The first meta-analysis included only 1489 patients from 11 trials, including 6 randomized controlled trials [8]. Recently three more meta-analyses of randomized controlled trials of SDD have been published [1–3]. These meta-analyses differ not only in the number of trials and patients analyzed but also in the methodology used. Kollef analyzed only published trials, including 2270 patients in 21 trials [3]. Heyland et al. analyzed 3395 patients in 25 randomized published and unpublished trials [2]. The SDD Trialists Collaborative Group analyzed 4142 patients from 22 randomized trials, both published and unpublished. They collected all patient data through direct contact with the investigators so that they could carry out the analysis on an intention to treat basis [1].

All meta-analyses show a significant reduction in the relative risk of respiratory infections in patients receiving intensive care. Kollef [3] found a highly significant reduction in pneumonia rate but a small and nonsignificant reduction in overall mortality (268 deaths out of 1105 for SDD-treated patients and 305 of 1165 control patients; $p=0.291$) [3].

The SDD Trialists' Collaborative Group found that SDD significantly reduced respiratory tract infections (odds ratio 0.37; 95% confidence interval

0.31–0.43). The value of the common odds ratio for total mortality (odds ratio 0.90; 95% CI 0.79–1.04) suggested a moderate treatment effect, reaching statistical significance only when the subgroup of trials of topical and systemic treatment combined was considered separately (odds ratio 0.80; 95% CI 0.67–0.97) [1].

Heyland et al. found that the incidence of pneumonia was significantly lower in patients receiving SDD RR 0.46; 95% CI 0.39–0.56), and that overall mortality was significantly reduced (RR 0.87; 95% CI 0.79–0.97). In subgroup analyses the largest absolute difference was found between studies using regimens including intravenous antibiotics (RR 0.80; CI 0.68–0.95) and those excluding systemic antibiotics (RR 0.99; CI 0.83–1.19) [2].

Effectiveness of Decontamination

The fact that the subgroup of studies using topical antibiotics in combination with intravenous prophylaxis showed a mortality benefit suggests that the systemic component is very important. However, this subgroup also differed in the decontamination regimen used. There are 14 studies that used intravenous antibiotic prophylaxis, of which 9 used oral and intestinal decontamination with the PTA regimen. In only two of the 14 studies was no decontamination achieved, due to improper choice of topical antibiotics, and the findings of these two studies were negative (Table 1). A total of 9 studies excluded intravenous antibiotic prophylaxis and found no reduction in mortality (Table 1). In only two of these studies was decontamination of both throat and gut performed. The remaining seven applied topical antibiotics either in the GI tract or only in the oral cavity; furthermore, many different antibiotic regimens were used without presenting any evidence of their effectiveness with regard to decontamination. Thus the difference in mortality between the two groups cannot be explained only by the use of the systemic component but also by the effectiveness of the topical regimen.

The fact that the effectiveness of decontamination is a crucial factor has been demonstrated in a recent study by Tetteroo et al. [4] Patients treated with SDD were divided into two groups: those who were effectively decontaminated within 3 days and those who were not. The actual death rate in patients who were effectively decontaminated within 3 days ($n=72$) was significantly lower than the predicted death rate (as calculated by the APACHE II score; 18% versus 40%, $p=0.006$), whereas no difference was found in those patients who were not effectively decontaminated despite the treatment with topical antibiotics ($n=25$, death rate 44%). The unsuccessful decontaminated patients had a significantly longer hospital (52 vs. 34 days) and ICU length of stay (23 vs. 9 days; $p=0.002$) and higher mortality rates (44% vs. 18%, $p=0.20$) when than the successfully decontaminated patients.

Table 1. Trials using topical antibiotics and parenteral prophylaxis

References	*n*	Topical regimen	Topical application	Systemic prophylaxis
Topical antibodies + parenteral prophylaxis				
Verhaegen [27]	268	Ofloxa	Oral + intestin	Ofloxa
Cockerill [28]	150	P, Gen, Nys	Oral + intestin	Ctx
Aerdts [29]	88	P, Nor, A	Oral + intestin	Ctx
Ulrich [30]	112	P, Nor, A	Oral + intestin	Tmp
Verhaegen [27]	385	½P, TA	Oral + intestin	Ctx
Rocha [24]	101	PTA	Oral + intestin	Ctx
Palomar [31]	97	PTA	Oral + intestin	Ctx
Blair [32]	231	PTA	Oral + intestin	Ctx
Kerver [33]	96	PTA	Oral + intestin	Ctx
Hammond [11]	322	PTA	Oral + intestin	Ctx
Ferrer [10]	80	PTA	Oral + intestin	Ctx
Jacobs [34]	91	PTA	Oral + intestin	Ctx
Sanchez [35]	271	PTA	Oral + intestin	Ctx
Winter [36]	183	PTA	Oral + intestin	Czd
Topical antibiotics alone				
Korinek [37]	191	PTA	Oral + intestin	None
Gastinne [14]	445	PTA	Oral + intestin	None
Cerra [38]	48	Nor, Nys	Oral + intestin	None
Godard [39]	185	PT	Intestin	None
Unertl [40]	39	P, Gen, A	Intestin	None
Gaussorgues [41]	118	½P, Gen, Van, A	Intestin	None
Brun Buisson [9]	133	½P, Neo, Nali	Intestin	None
Pugin [13]	79	P, Neo, Van	Oral	None
Rodriguez [42]	31	PTA	Oral	None

Ofloxa, ofloxacine; P, polymyxin/colistin; Gen, gentamicin; Nys, nystatin; Ctx, cefotaxime; Nor, norfloxacin; A, amphotericin B; Tmp, trimetoprim; ½P, half dose of polymyxin; Czd, ceftazidime; Neo, neomycin; Van, vancomycin; Nali, nalidixic acid.

Criteria for Pneumonia

In the control groups of SDD studies there is a strikingly wide range in the incidence of respiratory tract infections (18%–80%) although the criteria for diagnosis were very similar. This may be explained by the differences in patient populations and in underlying disease, but it has also been argued that due to nonspecific criteria used for diagnosis of pneumonia in unblinded studies the true pneumonia rate has been overestimated in SDD studies [9]. In fact, many of the respiratory tract infections may represent bronchopneumonia, bronchitis, or tracheitis rather than pneumonia. However, upper airway infections cannot be disregarded as trivial and without clinical importance.

There is little doubt that in studies in which pneumonia is diagnosed by the "protected brush technique" using a cutoff point of 10^3 CFU or by bronchoalveolar lavage with a cutoff point of 10^5 CFU/ml, the infection rate is much lower than in studies using clinical criteria. However, the SDD Trialists' Collaborative Group demonstrated that the reduction in relative pneumonia risk by SDD is independent of the criteria used for diagnosing pneumonia and independent of the blinding of the study. In the studies using the "protected brush" or bronchoalveolar lavage to diagnose pneumonia the odds ratio was 0.38 (CI 0.29–0.49) versus 0.36 (0.29–0.44) in the studies not using the protected brush [1].

Negative SDD Trials

A few trials using the full SDD scheme, including oral and intestinal decontamination with the PTA regimen and systemic antibiotic prophylaxis, did not find a significant reduction in the infection rate [10, 11]. There are several possible explanations for these negative results. One is the impact of SDD on the ecology of the ICU by eliminating oral and intestinal carriage of PPM. A recent controlled trial showed that the control group inside the same ICU as the SDD-treated group had a significantly lower acquisition, colonization, and infection rates than a concurrent control outside the ICU or than a control group in the same ICU after discontinuation of the SDD [12]. This demonstrated that a randomized controlled study conducted inside a single ICU as the "gold standard" may (partially) mask the effect of decontamination. Secondly, a number of double-blind studies have been conducted in an environment in which MRSA was endemic. It is obvious that MRSA, which is also resistant to the tobramycin in the PTA regimen, may be selected by SDD if no appropriate measures are taken, such as adding vancomycin to the regimen [13]. In these studies the omission of the surveillance cultures or the blinding to its results negatively influenced the results of the study due to colonization and infection with MRSA [10, 11, 14]. Thirdly, in some studies the problem of exogenous infections appears not to be solved yet given the high incidence of exogenous gram-negative pulmonary infections in both treated and untreated patients [11].

Preoperative SDD in High-Risk Surgery, Fulminant Liver Failure, and Burns

Meta-analyses have included only randomized trials in a general ICU population. Other trials with *preoperative* SDD have shown that it is highly effective in reducing the perioperative infection risk in high-risk surgery such as esophageal resections [15] and liver transplantation [16–18]. Other studies have shown that SDD reduces infections and mortality in patients with acute fulminant liver failure who are waiting for transplantation [19, 20]. The advantage of preoperative SDD is that intestinal decontamination can be achieved within 48 h because the bowels are still functioning, whereas when it is started postoperatively it may take 7–10 days before complete decontamination is achieved. Two retrospective of studies in burn patients show that SDD is effective in preventing infectious complications by intestinal enterobacteria and pseudomonas [21, 22].

Emergence of Resistance

Few controlled trials have discovered any evidence for the emergence of resistance. However, a recent study by Rocha [23] reported that cefotaxime and tobramycin resistance was greater in the SDD-treated group. In this study the copy strains (i.e., the same micro-organism from the same site from the same patient) were not excluded. Moreover, the authors observed a decrease in the absolute number of resistant strains rather than an increase; for example, tobramycin-resistant *Acinetobacter* was found in 47/143 (33%) isolates in the control group and in 14/23 (61%) of the treated group. Another report by Nardi et al. [24] showed an increase in tobramycin-resistant *Pseudomonas* isolates during 2 years of SDD compared to a historical control group; here the copy-strains were excluded and the percentages of isolates were reported rather than the prevalence of the resistant micro-organisms in the ICU. After 10 years of routine use of the same antibiotic regimen for selective decontamination in the same ICU there is still no evidence for emergence of resistance.

Selection of Gram-Positive Cocci

It has been suggested that SDD may lead to a gram-positive problem. It is obvious that enterococci and coagulase negative staphylococci, which belong to the normal indigenous flora, are the only micro-organisms isolated during successful decontamination. This does not represent overgrowth. Controlled trials that specifically studied enterococci and coagulase negative staphylococci have not shown an increased colonization or infection with these micro-organisms during SDD [11, 26].

As pointed out above, the PTA regimen is not effective against MRSA, and therefore this micro-organism may be selected [10, 11, 14, 27]. Whenever SDD is being used in an environment in which MRSA is endemic, vancomycin should be added to the PTA regimen [13]. Routine surveillance cultures to check the effectiveness of the decontamination remain the hallmark of SDD.

Conclusion

SDD is an infection prevention technique targeting primary and secondary endogenous infections by aerobic gram-negative bacteria and yeasts. The technique includes three elements: (a) oral and intestinal decontamination with topical nonabsorbable antibiotics, (b) a short-term systemic antibiotic prophylaxis, and (c) microbiological surveillance.

Meta-analyses have shown that SDD is highly effective in preventing respiratory tract infections in high-risk patients. It is more effective than any other published preventive measure. Exogenous infections are still important but can only be prevented with conventional hygienic measures. In an environment in which MRSA is endemic the regimen of topical antibiotics should be adjusted to avoid selection of MRSA. Even in the most successful SDD trials there is a discrepancy between the dramatic reduction in infection rate and the modest reduction in mortality. This clearly demonstrates that the underlying disease is a more important determinant of mortality than acquired infections. However, when SDD is fully implemented i.e., both oral and intestinal decontamination with the PTA regimen in combination with a short-term antibiotic prophylaxis and microbiological surveillance, a 20% reduction in mortality can be achieved.

The clinical relevance of the observed reduction in morality is apparent when it is compared with the few other interventions that have a proven effect on mortality. The effect of thrombolytic agents in acute myocardial infarction on mortality was demonstrated only after more than 14000 patients treated with thrombolytic agents were included in a cumulative meta-analysis. In none of the available studies did SDD lead to the emergence of resistance. However, bacteriological surveillance during SDD remains an essential part of the infection prevention technique. SDD is one of the best investigated and most effective infection prevention maneuvers in intensive care and should be considered in high-risk patients.

References

1. Selective Decontamination of the Digestive Tract Trialists' Collaborative Group (1993) Meta-analysis of randomised controlled trials of selective decontamination of the digestive tract. Br Med J 307:525–532
2. Heyland DK, Cook DJ, Jaeschke R, Griffith L, Lee HN, Guyatt GH (1994) Selective decontamination of the digestive tract: an overview. Chest 105:1221–1229
3. Koleff MH (1994) The role of selective digestive tract decontamination on mortality and respiratory tract infections. A meta-analysis. Chest 105:1101–1108
4. Tetteroo GWM, Wagenvoort JHT, Mulder PGH, Ince C, Bruining HA (1993) Decreased mortality rate and length of stay in surgical intensive care unit patients with successful selective decontamination of the gut. Crit Care Med 21:1692–1698
5. Stoutenbeek CP, van Saene HKF, Miranda DR, Zandstra DF (1984) The effect of selective decontamination of the digestive tract on colonization and infection rate in multiple trauma patients. Intensive Care Med 10:185–192
6. Feron B, Adair C, Gorman SP, McClurg B (1993) Interaction of sucralfate with antibiotics used for selective decontamination of the gastrointestinal tract. Am J Hosp Pharm 50:2550–2553
7. Langer M, Cigada M, Mandelli M et al (1987) Early onset pneumonia: a multicenter study in intensive care units. Intensive Care Med 13:342–346

8. VandenBroucke-Grauls CMJE, VandenBroucke JP (1991) Effect of selective decontamination of the digestive tract on respiratory tract infections and mortality in the intensive care units. Lancet 338:859–862
9. Brun Buisson C (1994) Selective decontamination in critical care. Interpreting the synthesized evidence (editorial). Chest 105:978–980
10. Ferrer M, Torres A, Gonzàlez J, Puig de la Bellacasa J, Gatell JM, Jiménez MT (1992) Utility of selective digestive decontamination in a general population of mechanically ventilated patients. Am Rev Respir Dis 145:A112
11. Hammond JMJ, Potgieter PD, Saunders GL et al (1992) Double-blind study of selective decontamination of the digestive tract in intensive care. Lancet 340:5–9
12. Bonten MJM, Gaillard CA, Johanson WG et al (1994) Colonization in patients receiving and not-receiving topical antimicrobial prophylaxis. Am J Respir Gul Care Med 150:1332–1340
13. Pugin J, Auckenthaler R, Lew DP, Suter PM (1991) Oropharyngeal decontamination decreases incidence of ventilator associated pneumonia. JAMA 265:2704–2710
14. Gastinne H, Wolff M, Delatour F et al (1992) A controlled trial in intensive care units of selective decontamination of the digestive tract with nonabsorbable antibiotics. N Engl J Med 326:594–599
15. Tetteroo GWM, Wagenvoort JHT, Castelein AL, Tilanus HW, Ince C, Bruining HA (1990) Selective decontamination to reduce Gram-negative colonization and infection after oesophageal resection. Lancet 335:704–707
16. Wiesner RH, Krom RAF, Hermans P (1988) Selective bowel decontamination to decrease gram-negative aerobic bacterial and candida colonization and prevent infection after orthotopic liver transplantation. Transplantation 45:570–574
17. Bion JF, Badger I, Crosby HA et al (1994) Selective decontamination of the digestive tract reduces Gram-negative pulmonary colonization but not sytemic endotoxemia in patients undergoing elective liver transplantation. Crit Care Med 22:40–49
18. Smith SD, Jackson RJ, Hannakan CJ, Wadowsky RM, Tzakis AG, Rowe MI (1993) Selective decontamination in pediatric liver transplants.Transplantation 55:1306–1309
19. Rolando N, Gimson A, Wade J, Philpott-Howard J, Casewell M, Williams R (1993) Prospective controlled trial of selective parenteral and enteral antimicrobial regimen in fulminant liver failure. Hepatology 17:196–201
20. Salmeron JM, Tití L, Rimola A et al (1992) selective decontamination of the digestive tract in the prevention of bacterial infection in patients with acute liver failure. J Hepatol 14:280–285
21. Mackie DP, van Hertum WAJ, Schumburg T, Kuijper EC, Knape P (1992) Prevention of infection in burns: preliminary experience with selective decontamination of the digestive tract in patients with extensive injuries. J Trauma 32:570–575
22. Manson WL, Pernot PCJ, Fidler V, Sauer EW, Klasen HJ (1992) Colonization of burns and the duration of hospital stay of severely burned patients. J Hosp Infect 22:55–63
23. Rocha L A, Martin MJ, Pita S et al (1992) Prevention of nosocomial infection in critically ill patients by selective decontamination of the digestive tract. A randomized double blind, placebo controlled study. Intensive Care Med 18:398–404
24. Nardi G, Valentis U, Proietti A et al (1993) Epidemiological impact of prolonged systematic use of topical SDD on bacterial colonization of the tracheobronchial tree and antibiotic resistance. Intensive Care Med 19:273–288
25. Hammond JMJ, Potgieter PD (1995) Long-term effects of selective decontamination on antimicrobial resistance. Gul Care Med 23:637–645
26. Humphreys H, Winter R, Pick A (1992) The effect of selective decontamination of the digestive tract on gastrointestinal enterococcal colonization in ITU patients. Intensive Care Med 18:459–463
27. Verhaegen J (1992) Randomized study of selective digestive decontamination on colonisation and prevention of infections in mechanically ventilated patients in the ICU. Thesis, Leuven

28. Cockerill FR, Muller SM, Anhalt JP et al (1992) Prevention of infection in critically ill patients by selective decontamination of the digestive tract. Ann Intern Med 117:545–553
29. Aerdts SJ, van Dalen R, Clasener HAL, Festen J, van Lier HJJ, Vollaard EJ (1991) Antibiotic prophylaxis of respiratory tract infections in mechanically ventilated patients. A prospective, blinded, randomized trial of the effect of a novel regimen. Chest 100:783–791
30. Ulrich C, Harinck-de Weerd JE, Bakker NC, Jacz K, Doornbos L, de Ridder VA (1990) Selective decontamination of the digestive tract with norfloxacin in the prevention of ICU-acquired infections: a prospective randomized study. Intensive Care Med 15:424–431
31. Palomar M (1991) Prevention of nosocomial pneumonia in ventilated patients using cefotaxime (CTX) and SDD. 17th international congress on chemotherapy, Berlin
32. Blair P, Rowlands BJ, Lowry K, Webb H, Armstrong P, Smilie J (1991) Selective decontamination of the digestive tract: a stratified, randomized, prospective study in a mixed intensive care unit. Surgery 110:303–310
33. Kerver AJH, Rommes JH, Verhage EAE (1988) Prevention of colonization and subsequent infection in surgical intensive care patients. A prospective randomized study. Crit Care Med 16:1087–1093
34. Jacobs S, Foweraker JE, Roberts SE (1992) Effectiveness of selective decontamination of the digestive tract (SDD) in an ICU with a policy encouraging a low gastric pH. Clin Intensive Care 3:52–58
35. Sanchez M, Lopez J, Galvan B, Calvo C, Nuñez A, Cambronero JA, Alvarez J, Rogero S, Chamorro C, Rodriguez JM, Guerra L, Rubio J, Cerdá E (1990) Selective digestive tract decontamination (SDD) and sucralfate (S) in intubated patients. Intensive Care Med 16:S32 (abstract)
36. Winter R, Humphreys H, Pick A, MacGowan AP, Willatts SM, Speller DCE (1992) A controlled trial of selective decontamination of the digestive tract in intensive care and its effect on nosocomial infection. J Antimicrob Chemother 30:73–87
37. Korinek AM, Laisne MJ, Nicolas MH, Raskine L, Deroin V, Sanson-Lepors MJ (1993) Selective decontamination of the digestive tract in neurosurgical intensive care unit patients: a double blind, randomized, placebo-controlled study. Crit Care Med 21:1466–1473
38. Cerra FB, Maddaus MA, Dunn DL, Wells CL, Konstantinides NN, Lehmann SL, Mann HJ (1992) Selective gut decontamination reduces nosocomial infections and length of stay but not mortality or organ failure in surgical intensive care unit patients. Arch Surg 127:163–169
39. Godard J, Guillaume C, Reverdy ME, Bachmann P, Bui-Xuan B, Nageotte A, Motin J (1990) Intestinal decontamination in a polyvalent ICU: a double-blind study. Intensive Care Med 16:307–311
40. Unertl K, Ruckdeschel G, Selbmann HK, Jensen U, Forst H, Lenhardt FP, Peter K (1987) Prevention of colonization and respiratory infections in longterm ventilated patients by local antimicrobial prophylaxis. Intensive Care Med 13:106–113
41. Gaussorgues P, Salord F, Sirodot M, Tigaud S, Cagnin S, Gerard M, Robert D (1991) Efficacité de la décontamination digestive sur la survenue des bactériémies nosocomiales chez les patients sous ventilation mécanique et recevant des bêtamimétiques. Rean Soins Intens Med Urg 7:169–174
42. Rodriguez-Roldan JM, Altuna-Cuesta A, Lopez A, Carillo A, Garcia J, Leon J, Martinez-Pellus AJ (1990) Prevention of nosocomial lung infection in ventilated patients: use of an antimicrobial pharyngeal nonabsorbable paste. Crit Care Med 18:1239–1242

Effect of Aspirin on Central and Peripheral Cardiovascular Adaptation to Exercise in Heart Failure: Participation of Splanchnic Flow

E. Domingo, M. R. Gilabert, and R. Ballester

Introduction

Heart failure is acompanied by a variety of compensatory mechanisms (e.g., sympathetic nervous system, renin-angiotensin-aldosterone, prostaglandin axis, vasopressin, atrial natriuretic peptide) affecting regional as well as central hemodynamics. In general, studies in animals and humans suggest that in congestive heart failure there is a redistribution of regional blood flow away from vascular beds sensitive to sympathetic stimulation (e.g., splanchnic, renal) with a relative increase in the proportion of cardiac output to the more autoregulated vascular beds such as the coronary and cerebral circulations. Regional blood flow measurements in previous studies have been restricted to one organ system and most often in the resting condition. Studies analyzing the regional distribution of cardiac output during maximal exercise in chronic stable heart failure are lacking.

It has been known for at least 50 years that some prostaglandins decrease systemic blood pressure [1]. This pharmacological action by prostaglandins, well documented both in isolated preparation and in intact animals of various species, had not been convincingly shown to have a clearcut physiological equivalent until the discovery by Vane [2] that certain anti-inflamatory drugs inhibited the bioformation of prostaglandins. This tool allowed demonstration of the contribution of endogenous prostaglandins in the physiological regulatory mechanisms within the cardiovascular system. The effect of the prostaglandin axis, analyzed through its blockade with aspirin or indomethacin, on the regional distribution of blood flow in resting healthy subjects has been partially studied [3–6]. Within the setting of severe clinical heart failure the deleterious effect of prostaglandin blockade [7] and the increased plasma levels of degradation products of prostaglandin E_2 [8] have suggested that the prostaglandin axis is an operative mechanism in the circulatory homeostasis of patients with heart failure, at least when resting and supine. In human congestive heart failure there is a redistribution of regional blood flow away from vascular beds sensitive to the sympathetic nervous system stimulation (renal, splanchnic) to the predominantly autoregulated territories such as the coronary circulation [9]. The determination of regional blood flow responses in human heart failure will become more important as we expand our knowledge base of the pathophysiology of heart failure.

Since heart failure displays, in addition to a decreased ventricular pumping efficacy, an abnormal distribution of regional blood flows in both the resting and exercise conditions, and since the prostaglandin axis seems to be highly stimulated in heart failure, we attempted to determine whether the inhibition of prostaglandin production by aspirin would alter the central and peripheral hemodynamic adaptation to upright exercise in both healthy subjects and moderate heart failure patients.

In congestive heart failure blood flow to hepatic, renal, and limb regions is significantly decreased at rest, and its decrease is linearly related to the reduction in cardiac output. Considerable individual variation in regional flow and in regional vascular resistance has observed in normal subjects and in patients with congestive heart failure.

The splanchnic circulation is a system of different organs in series and parallel including the gut, spleen, liver, and pancreas. Since all its flow drains into the hepatic vein, and since most of the techniques used for its assessment measure global changes and are therefore unable to discriminate individual organ variations, splanchnic flow is usually considered hypothetically as a single functional unit, despite the nonhomogeneous behavior of these organs (particularly during digestion).

In healthy individuals splanchnic flow is 25% of cardiac output in the supine and resting state, aproximately 1500 ml/min. In heart failure, depending on its severity, it ranges from 500 to 1000 ml/min in the supine position. The percentage of volemia stored in the splachnic circulation is 20–25%. Its regulation in the healthy subject is mainly through the autonomic sympathetic nervous system. In healthy subjects during maximal exercise splanchnic flow may decrease up to five times.

Material and Methods

We prospectively studied ten healthy normotensive male volunteers with a normal two dimensional echo-Doppler study, who were not included in any regular program of exercise and were aged 28–45 years, and ten patients aged 35–57 years with moderate chronic congestive heart failure class II (New York Heart Association classification). All subjects were normotensive white men. They followed a liberal diet throughout the study and stopped all medications 1 week before entering the study. Patients with heart failure had no clinical symptoms of ischemic heart disease nor electrocardiogram changes suggesting exercise ischemia. Their echocardiographic-Doppler study showed the absence of any significant mitral and aortic valvular regurgitation and the absence of ventricular diastolic dysfunction. All heart failure patients had left ventricular dilatation, with echocardiographic left ventricular diameters ranging from 58 to 69 mm. All subjects gave their informed consent and were familiar to bycicle stress testing. All healthy subjects and patients were submitted to an exercise test before entering the study to determine the maximal workload which they were able to carry out.

The exercise was undertaken on an upright bycicle ergometer in the sitting position (Siemens 740). The exercise protocol consisted of discontinous periods of exercise of 5 min duration. The external workload was kept constant for each exercise period. There was a resting period of 5 min between each exercise burst. Healthy subjects exercised initially at 50 W, and then the external workload was increased to the level previously determined for all subjects in order to reach as close as possible their subjective maximal exercise capacity, not precluding the measurement of all variables during the last 30 s of each exercise period. Healthy subjects reached a maximal workload between 150–200 W. All heart failure patients reached their exhaustion at the end of the initial 50 W exercise period. Environmental conditions were carefully controlled, monitoring room temperature at 22and room air humidity at 40%–60%.

All exercise protocols were performed randomly and singly blinded by all subjects under placebo and under oral aspirin between 14–16 p.m. to avoid the influence of circadian rhythm on the cardiovascular adaptative mechanisms to exercise.

Aspirin dosage was fixed at 500 mg orally every 6 h the last 24 h before the study. The exercise test was undertaken 2–4 h after the last dose. The studies carried out under placebo were taken as control.

Tracings were measured, and variables derived from these measurements were calculated blindly, without knowledge of placebo and aspirin conditions. Variables were simultaneously obtained in the sitting position under resting condition and during the last 30 s of any level of exercise. Resting condition was defined as the stable condition (three or more similar consecutive values of heart rate and blood pressure at 30-s intervals) after at least 10 min of sitting quietly on the bycicle. Stroke volume, cardiac output, forearm blod flow, leg blood flow, and splanchnic blood flow were measured simultaneously and noninvasively by the tetrapolar electrical impedance technique (Minnesota Impedance Cardiograph, MIC 304A [10, 11]). Heart rate was monitored by the electrogram. Blood pressure was measured noninvasively by an automatic pressure cuff (BP-103 N Mark III) every 30 s. Total and peripheral regional resistances were calculated as the ratio of mean blood pressure/blood flow, ignoring right atrial pressure. Normal or near-normal right atrial pressure was assumed in heart failure patients since neither at rest nor at peak exercise were signs of right ventricular failure detected. Cardiac output was obtained in liters per minute, blood pressure in mmHg, total peripheral resistance in mmHg per liter per minute, arm blood flow and leg blood flow in milliliters per minute per 100 ml tissue, arm, leg and splanchnic resistances in mmHg per milliliter per minute per 100 ml tissue. Mean blood pressure was calculated as diastolic blood pressure plus 1/3 pulse pressure.

Statistical comparison of healthy subjects under placebo versus healthy subjects under aspirin, and heart failure patients under placebo versus heart failure patients under aspirin used Student's *t* test for paired data, since each subject was his own control. Comparison of healthy subjects versus heart failure patients used Student's t test for unpaired data. Statistical significance was defined as $p < 0.05$.

Results

Tables 1 and 2 present the results of both groups of subjects under all levels of studied exercise intensity and under either placebo and aspirin.

Resting Condition

Nonsignificant differences between placebo and aspirin were found in most of the variables at rest both in healthy and in heart failure subjects. However, aspirin induced a negative chronotropic effect in both groups ($p < 0.05$; Fig. 1). In addition, aspirin was associated in the healthy population with a tendency to vasoconstriction ($p = 0.09$) and to a decreased flow ($p = 0.06$) in the leg. Heart failure patients had higher heart rate (Fig. 1), systemic vasoconstriction, and diastolic blood pressure ($p < 0.05$; Figs. 2, 3) but similar resting cardiac output and systolic blood pressure as healthy subjects.

Table 1. Peripheral parameters under placebo and aspirin (ASA) in healthy and in heart failure subjects

	Leg		Arm	
	Blood flow	Peripheral resistance	Blood flow	Peripheral resistance
Healthy subjects				
Rest				
Control	3.0 ± 1.0	29.6 ± 9.5	2.6 ± 0.7	32.9 ± 9.9
ASA	2.4 ± 0.4	34.5 ± 8.3	2.8 ± 1.0	28.9 ± 9.1
Exercise 50 W				
Control	4.5 ± 1.2	20.5 ± 6.4	3.8 ± 1.0	24.9 ± 8.7
ASA	3.7 ± 0.6	22.7 ± 4.8	3.4 ± 0.9	27.3 ± 8.5
Exercise max.				
Control	13.6 ± 5.0	8.4 ± 2.3	8.0 ± 1.1	13.5 ± 2.5
ASA	19.0 ± 13.0	8.9 ± 4.3	9.2 ± 4.3	14.7 ± 4.2
Heart failure subjects				
Rest				
Control	3.4 ± 1.5	29.6 ± 12.4	4.1 ± 0.8	21.3 ± 4.4
ASA	4.6 ± 3.7	28.1 ± 10.3	4.1 ± 1.1	22.7 ± 7.3
Exercise 50 W				
Control	6.7 ± 3.2	20.1 ± 11.5	5.7 ± 1.3	19.9 ± 3.5
ASA	6.2 ± 27	21.1 ± 9.3	6.1 ± 0.9	18.8 ± 2.4

Table 2. Central hemodynamic parameters under placebo and aspirin (ASA) in healthy and in heart failure subjects

	Heart rate	Cardiac output	Blood pressure (mean)	Total peripheral resistance
Healthy subjects				
Rest				
Control	66.4 ± 5.4	4.6 ± 0.9	79.6 ± 7.8	17.8 ± 3.8
ASA	62.1 ± 4.8	4.4 ± 0.7	78.3 ± 7.1	18.5 ± 3.5
Exercise 50 W				
Control	87.5 ± 7.0	7.8 ± 1.6	82.0 ± 10.0	11.1 ± 3.4
ASA	88.5 ± 8.8	8.8 ± 2.4	87.1 ± 11.0	10.0 ± 3.5
Exercise max.				
Control	181 ± 11	21.2 ± 3.5	106 ± 8.7	5.0 ± 0.8
ASA	175 ± 15	21.9 ± 3.5	125 ± 4.0	5.8 ± 0.9
Heart failure subjects				
Rest				
Control	85 ± 13.4	4.2 ± 0.8	85 ± 5.8	20.8 ± 4.6
ASA	76 ± 10.6	4.0 ± 1.0	87 ± 5.8	22.7 ± 7.0
Exercise 50 W				
Control	159 ± 13.0	9.3 ± 1.9	113 ± 10.7	12.5 ± 3.1
ASA	151 ± 14.0	9.1 ± 2.2	112 ± 10.6	13.5 ± 6.5

Exercise Condition

Heart Rate. Aspirin had a more marked chronotropic effect than placebo under moderate exercise in healthy subjects ($p < 0.04$). It did not change its behavior in the other studied conditions. It must be pointed out that heart failure subjects reached their maximal exercise capacity with the first workload of 50 W, as shown by their heart rate at peak exercise (Fig. 1).

Systolic Blood Pressure. Systolic blood pressure during maximal exercise in heart failure patients reached absolute values between those reached under moderate and maximal exercise by healthy subjects. In healthy subjects aspirin caused higher increments in systolic blood pressure ($p = 0.04$) at maximal exercise, but in heart failure subjects the opposite tendency was found ($p = 0.08$; Fig. 2).

Diastolic Blood Pressure. In heart failure patients no differences in the behavior of diastolic blood pressure were caused by aspirin. In healthy subjects similar qualitative effects were induced by aspirin at both exercise levels: a higher increment under maximal exercise ($p = 0.03$) and no decrement under moderate exercise. (Fig. 2).

Cardiac Output. Cardiac output increased significantly for all subjects of both groups under all exercise intensities ($p < 0.05$). However, heart failure

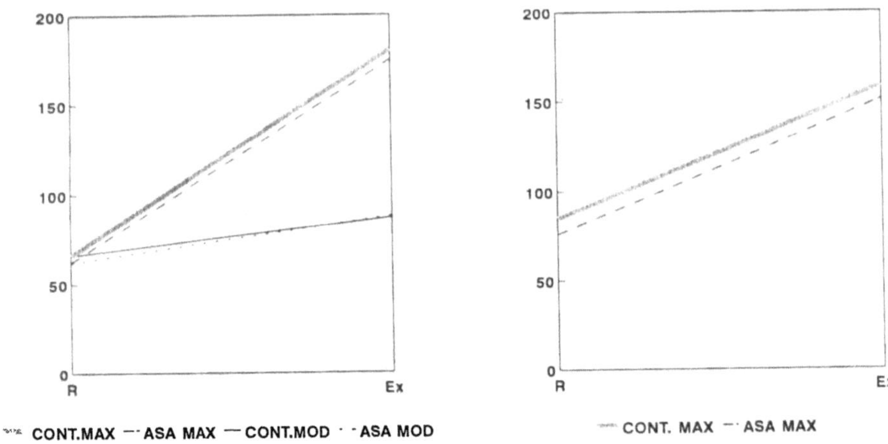

Fig. 1. Mean heart rate among healthy subjects (*left*) and heart failure patients (*right*) under sitting resting conditions (*R*) and under exercise (*EX*). *Cont*, control; *ASA*, aspirin; *Mod*, moderate; *Max*, maximal

patients at their maximal exercise reached a similar cardiac output to that of healthy subjects at moderate exercise (Fig. 3). In healthy subjects but not in heart failure patients aspirin caused a tendency to higher increments in cardiac output ($p = 0.08$) only at moderate exercise.

Total Peripheral Resistance. In healthy subjects under moderate exercise aspirin induced higher but nonsignificant decrements. Aspirin did not change its behavior either in healthy or in heart failure subjects at maximal exercise capacity (Fig. 3).

Arm Hemodynamics. In healthy subjects under moderate exercise aspirin caused a nonsignificant minor increased blood flow and vasodilatation compared to placebo, but it did not change the arm hemodynamics at maximal exercise. However, in heart failure patients significant differences were found: aspirin caused a major blood flow increment ($p = 0.01$) and vasodilatation ($p = 0.02$; Fig. 4).

Leg Hemodynamics. Nonsignificant changes were induced by aspirin on leg hemodynamics during both exercise intensities in healthy subjects. However, in heart failure patients aspirin caused a minor increment in blood flow ($p = 0.04$) and minor vasodilatation ($p = 0.02$) with exercise (Fig. 5).

The direction of qualitative changes on all variables under all studied conditions in both groups and their statistical significance are summarized graphically in Fig. 6.

Splanchnic Circulation. At maximal exercise in heart failure patients, splachnic flow increased proportionally more under aspirin ($p < 0.05$), reaching ab-

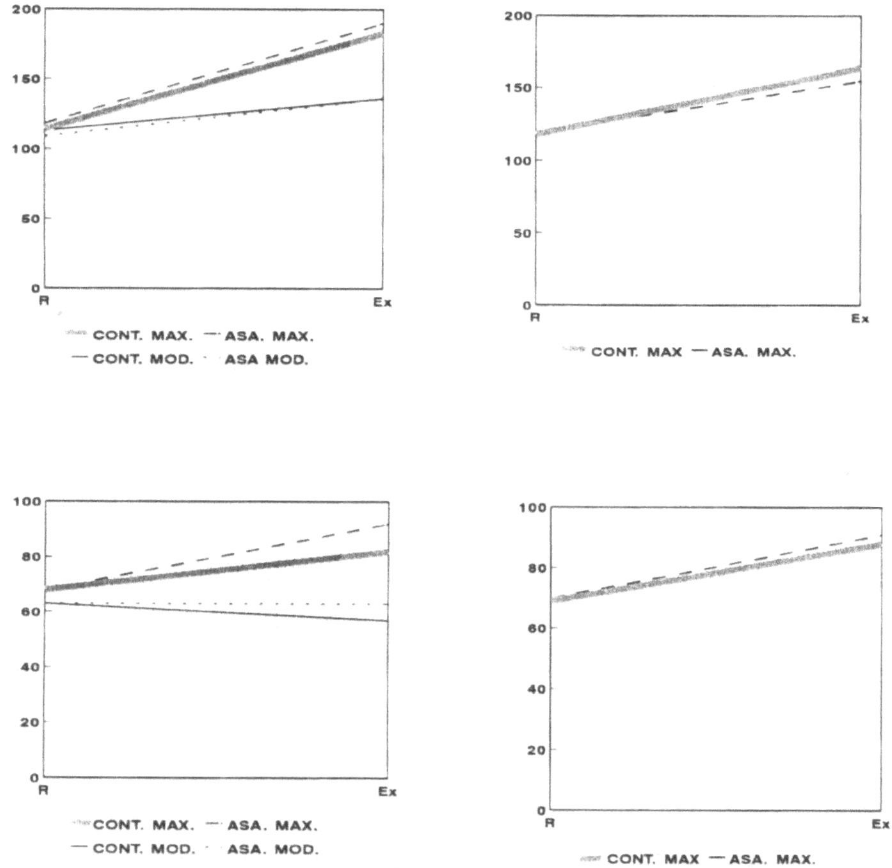

Fig. 2. Mean systolic (*above*) and diastolic (*below*) blood pressure among healthy subjects (*left*) and heart failure patients (*right*) under sitting resting conditions (*R*) and under exercise (*EX*). *Cont*, control; *ASA*, aspirin; *Mod*, moderate; *Max*, maximal

solute values closed to those under placebo, although rising from lower resting values. Splanchnic resistance did not change during exercise under aspirin, while under placebo it increased, reaching absolute values similar to those under aspirin although starting from a higher basal vasodilatation ($p<0.05$). The splanchnic impedance increased similarly during exercise in both conditions, suggesting an increased venous blood pooling effect under aspirin (Fig. 7).

Discussion

In recent years cardiologists have developed three different pathophysiological models of the heart failure condition: the cardiorenal, cardiovascular,

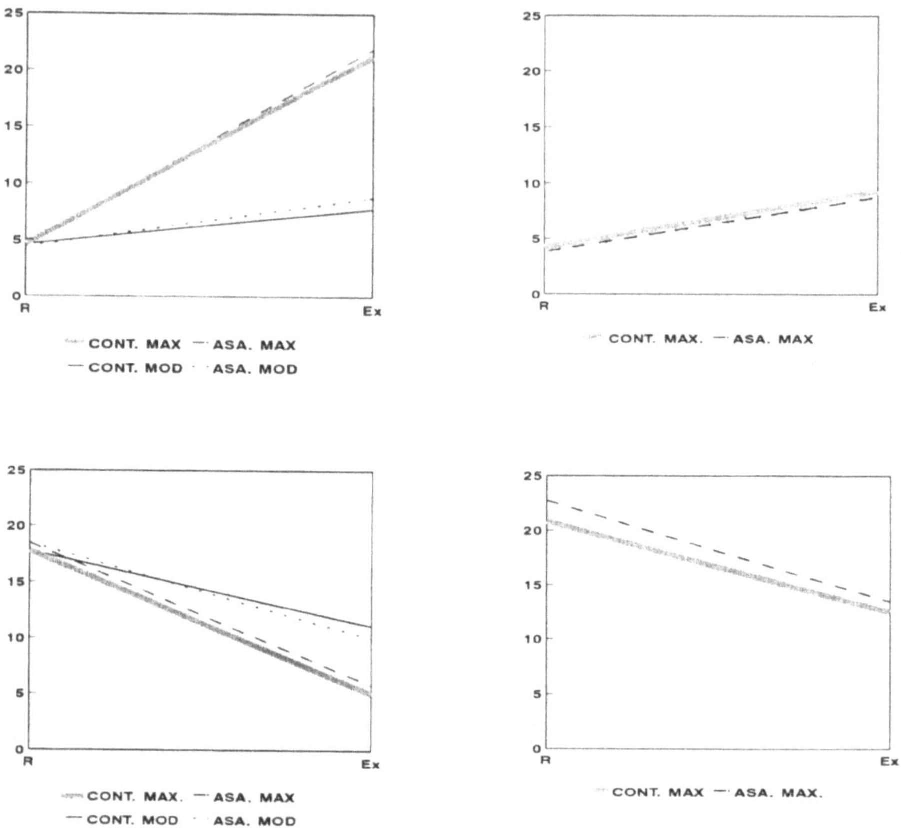

Fig. 3. Mean cardiac output (*above*) and total peripheral resistance (*below*) among healthy subjects (*left*) and heart failure patients (*right*) under sitting resting conditions (*R*) and under exercise (*EX*). *Cont*, control; *ASA*, aspirin; *Mod*, moderate; *Max*, maximal

and neurohormonal models [12]. The prostaglandin axis is probably involved in the formulation of all three models.

Despite the total body consequences of congestive heart failure little information is available on the distribution of cardiac output and regional organ blood flow in this condition in humans, probably due to technical and methodological limitations in the simultaneous and repetitive monitoring of regional flows under dynamic conditions. Our results do not fully agree with those of other published studies on the effect of the pharmacological prostaglandin axis blockade in heart failure [13]. These studies were performed mostly at rest, in more severe heart failure, and using another prostaglandin synthetic blocking drug such as indomethacin.

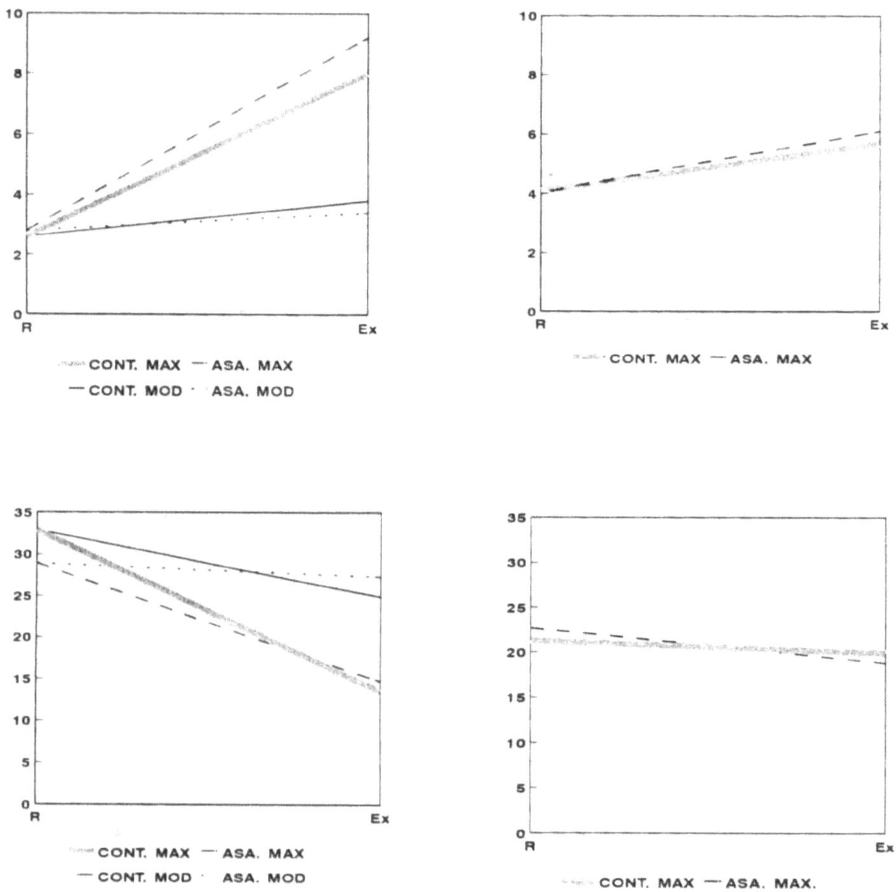

Fig. 4. Mean forearm blood flow (*above*) and forearm peripheral resistance (*below*) among healthy subjects (*left*) and heart failure patients (*right*) under sitting resting conditions (*R*) and under exercise (*EX*). *Cont*, control; *ASA*, aspirin; *Mod*, moderate; *Max*, maximal

Heart Rate

As expected, heart rate increased with exercise proportionally to its intensity. Aspirin did not induce significant changes in the heart rate adaptation to exercise either in heart failure in healthy subjects under maximal exercise (Fig. 1, Tables 1, 2). However, in both groups the resting heart rate was lower than placebo, suggesting a negative chronotropic effect of prostaglandin axis blockade. It has previously been reported in animal studies under resting condition that prostaglandins may induce a positive chronotropic effect [14], in accordance with our results where prostaglandins blockade was related to a negative chronotropic effect. However, under moderate exercise healthy subjects had a greater increase in heart rate under aspirin, probably

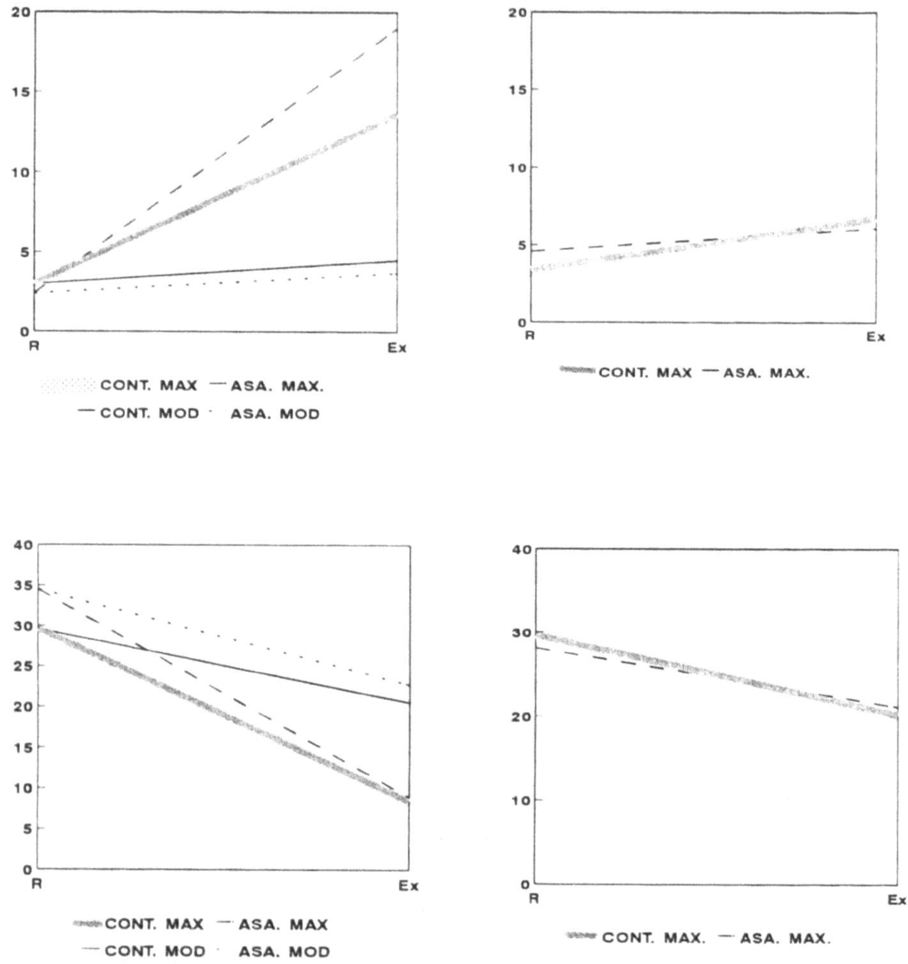

Fig. 5. Mean leg blood flow (*above*) and leg peripheral resistance (left) among healthy subjects and heart failure patients (*right*) under sitting resting condition (*R*) and under exercise (*EX*). *Cont*, control; *ASA*, aspirin; *Mod*, moderate; *Max*, maximal

an indirect and reflex effect due to the greater systemic vasodilatation (Fig. 3) or merely the response to starting from lower resting heart rate levels, since the reached heart rate was similar.

Blood Pressure

Diastolic blood pressure was not influenced by aspirin in heart failure patients either at rest or during exercise, suggesting no influence on total systemic resistance (Fig. 2, Table 2). However, in healthy subjects aspirin was

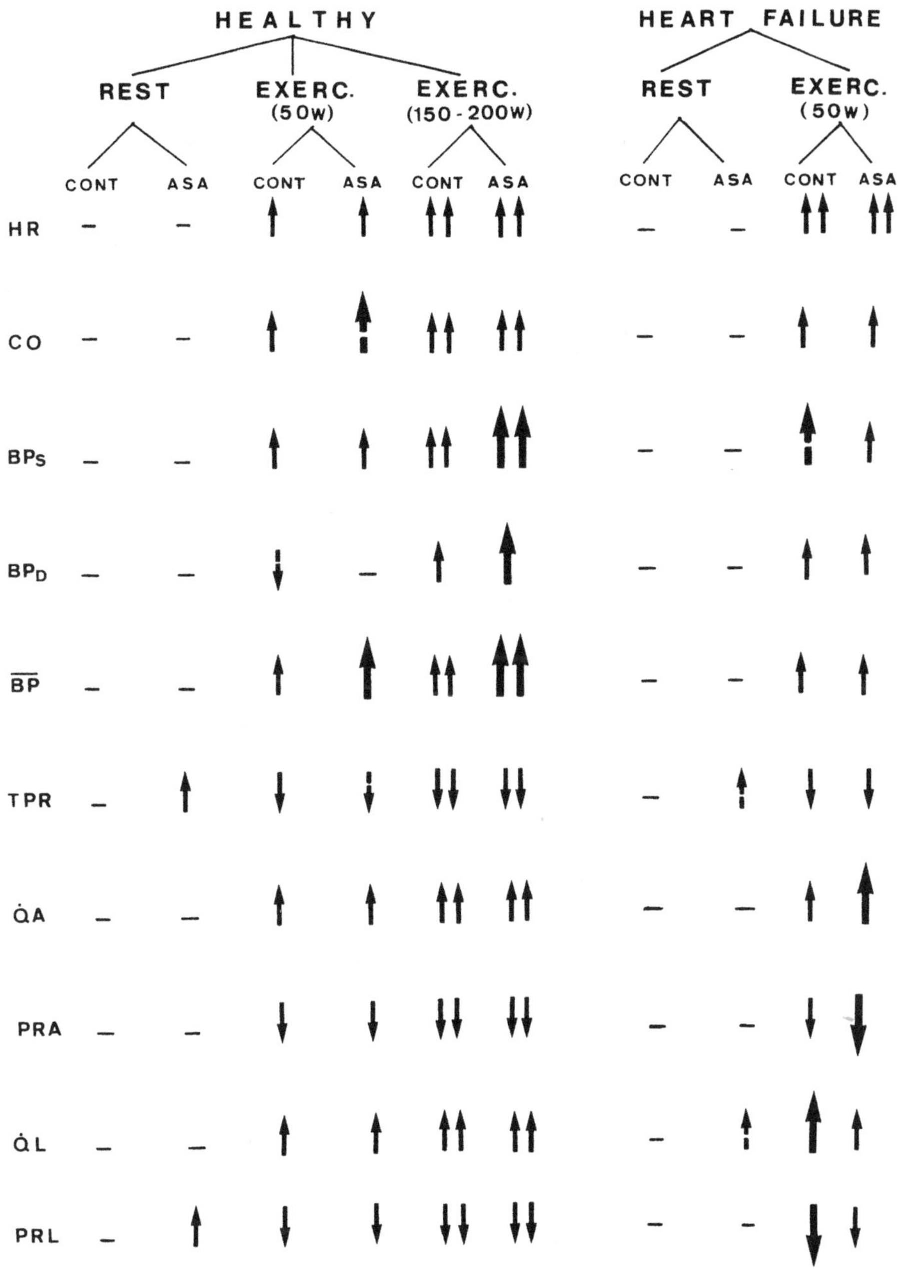

Fig. 6. Summary of qualitative behavior and statistical significance of the studied variables under all conditions in both groups of subjects. *Cont*, control; *ASA*, aspirin; *Mod*, moderate; *Max*, maximal. *Magnitude of arrow*, statistically significant changes between aspirin and control under the same condition ($p<0.05$); *number of arrows*, magnitude of the absolute change of the variable; *discontinued arrows*, a nonsignificant tendency ($0.05<p<0.10$)

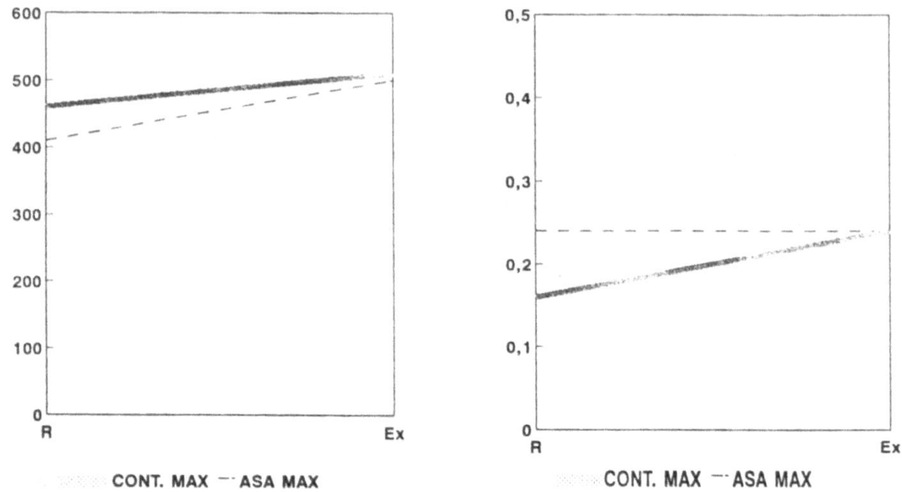

Fig. 7. Mean splanchnic blood flow (*left*) and splanchnic peripheral resistance in heart failure patients under sitting resting condition (*R*) and under exercise (*EX*). *Cont*, control; *ASA*, aspirin; *Mod*, moderate; *Max*, maximal

associated at maximal exercise with a greater increase in diastolic blood pressure, related to a minor systemic vasodilatation, as expected when blocking a predominantly vasodilator axis such as prostaglandin (Fig. 4, Table 1). This behavior was not found in healthy subjects at moderate exercise intensity, a paradoxical effect probably related to a different influence of prostaglandins in the regulation of blood flow in different territories, such as splachnic flow [15, 16]. Systolic blood pressure showed paradoxical changes in healthy and in heart failure subjects since aspirin favored a tendency to a major increment in the former and a minor increment in the latter, probably in accordance with the cardiac output behavior.

Cardiac Output

In healthy subjects, as just mentioned, aspirin induced a higher increase in cardiac output during exercise, as reported in other studies using indomethacin as prostaglandin synthesis blocker [17]. Under moderate exercise this effect can be related to a hyperdynamic circulation (greater aspirin systemic vasodilatation) since reports on the inotropic effect of prostaglandin in animal experiments are contradictory, and human studies are lacking (Fig. 3, Table 1) [14]. However, in heart failure patients no changes were induced by aspirin in cardiac output behavior with exercise, suggesting that maximal cardiac output in the heart failure patients does not depend on the adaptation of the peripheral circulation, at least through the prostaglandin axis, since it may be limited by the left ventricular depressed pumping action (Fig. 3, Table 2). The electrical impedance technique is useful for estimating cardiac output. Even if the absolute values of cardiac output calculated by this technique may be 10%–15% higher than those obtained by invasive methods such as thermodilution or dye dilution, the resting values and the intraindividual variations are highly reproducible.

Total Peripheral Resistance

Both healthy subjects and heart failure patients showed some, albeit quantitatively different, vasodilatation during exercise. Heart failure patients at maximal exercise vasodilated similarly to healthy subjects at moderate exercise, suggesting an already reported impaired vasodilatation capacity in heart failure related to both a basal increased stimulation of vasodilators axis and to a decreased arterial wall compliance [18] (Fig. 3, Tables 1, 2). As expected, aspirin induced at rest, both in healthy and in heart failure subjects, a tendency to vasoconstriction. It did not affect the response of total peripheral resistance during exercise in heart failure subjects (Fig. 3, Table 2). However, even if total peripheral resistance behavior did not change under aspirin, what did change is the contribution of regional resistance to it. As explained below, this is part of the abnormal regional blood flow distribution induced by aspirin in heart failure patients during exercise, an abnormality that cannot be detected when systemic peripheral circulation is analyzed as a whole (Figs. 3–5, Table 2).

However, in healthy subjects aspirin induced a higher vasodilatation at moderate exercise and a minor vasodilatation at maximal exercise (Fig. 3, Table 2). This paradoxical effect of aspirin in healthy individuals at different levels of exercise may be explained by its influence on the splanchnic circulation [15, 19–21], since renal blood flow appears to be protected in human heart failure by a form of autoregulation, even during marked depression of cardiac output [22, 23]. It may well be that at moderate exercise, with a moderate sympathetic stimulation, healthy subjects showed increased splanchnic vasodilalation due to the blockade of vasoconstrictor splachnic prostaglandins [19–21], as shown in postprandial abdominal blood flow studies. However, at maximal exercise the effect of the more marked sympathetic vasoconstriction may completely mask the prostaglandin effect on the splanchnic circulation.

Arm and Leg Hemodynamics

In healthy subjects aspirin caused no change in blood flow or resistance adaptation to exercise in the exercising limb, that is, the leg, at any level of exercise (Fig. 5, Table 1). In heart failure patients aspirin decreased the blood flow and the degree of vasodilation in the leg at maximal exercise, thus favoring, again, a disregulation in the distribution of cardiac output during exercise (Fig. 5, Table 2). A recently published paper partially analyzing the response of cardiac output and forearm blood flow during supine exercise in heart failure suggested an increased participation of the prostaglandin axis in this condition [24].

In healthy persons in the nonexercising limb aspirin did not change the adaptation pattern to maximal exercise. However, at moderate exercise it did decrease the degree of vasodilatation to the arm, in spite of no change in the exercising limb, as noted above (Fig. 4, Table 1). In heart failure patients aspirin favored the distribution of cardiac output to the nonexercising limb (arm) and unfavored flow to the exercising limb (leg), thus contributing again to the disregulation of cardiac output distribution under maximal exercise in heart failure patients (Figs. 4, 5).

Splanchnic Hemodynamics

Heart failure induced a decreased splanchnic flow in our series and in the literature [25]. During exercise in heart failure patients blood flow did not decrease and even increased mildly and proportionally more under aspirin ($p=0.06$ vs. $p<0.05$; Fig. 7). This is probably due to the increased basal vasoconstriction even at rest, necessary to redistribute splanchnic blood flow to more vital territories. The increased venous blood pooling under aspirin during exercise (a higher increased of blood flow under aspirin and a simultaneous similar behavior under both conditions of splanchnic impedance, that is, of volume of blood stored) suggests a deleterious effect of the prostaglandin axis blockade, decreasing the contribution of splanchnic circulation to the increased venous return expected in a normal hemodynamic adaptation to exercise.

Conclusion

Our results suggest that in heart failure patients the correct regional organ blood flow distribution during exercise is much more prostaglandin axis dependent since aspirin administration induced a less efficient hemodynamic adaptation to exercise, favoring flow to nonworking territories. This is in accordance with the reported higher increments on all neurohumoral parameters, particularly metabolites of prostaglandin axis, during exercise in heart failure patients [24]

It must be pointed out that there is a paucity of really comparable data, since manuscripts on this subject usually report data under different conditions, calculating different variables and studying different populations. This it extremely difficult to carry out an integrated and definitive analysis of the role of the prostaglandin axis in central and peripheral hemodynamic adaptation to exercise in heart failure conditions. Nevertheless the present study indicates that a drug without direct hemodynamic activity, such as aspirin, can indeed induce changes in the hemodynamic behavior of the adaptation to exercise in heart failure conditions, even if these changes cannot be detected at rest. The clinical relevance of these changes remains to be determined.

References

1. Von Euler US (1936) On the specific vasodilating and plain muscle stimulating substances from accessory genital glands in man and certain animals (prostaglandin and vesiglandin). J Physiol (Lond) 88:213–234
2. Vane JR (1971) Inhibition of prostaglandin synthesis as a mechanism of action for aspirin-like drugs. Nature New Biol 231:235–237
3. Cowley AJ, Stainer K, Rowley JM, Hanley SP (1984) The effect of aspirin on peripheral haemodynamics changes following submaximal exercise in normal volunteers. Cardiovasc Res 18:511–513
4. Gerber JG (1983) Indomethacin-induced rises in blood pressure. Ann Intern Med 99:555–557
5. Cowley AJ, Stainer K, Rowley JM, Wilcox RG (1985) Effect of aspirin and indometahcin on exercise induced changes in blood pressure and limb blood flow in normal volunteers. Cardiovasc Res 19:177–180
6. Nowak J, Wennmalm A (1978) Influence of indomethacin and of prostaglandin E1 on total and regional blood flow in man. Acta Physiol Scand 102:484–491
7. Mc Carthy SJ, Jones VE, Romero SC (1982) Acute intrinsic renal failure induced by indomethacin. Role of prostaglandin synthetase inhibition. Mayo Clin Proc 57:289–296
8. Punzengruber C, Stanek B, Sinzinger H, Silberbauer K (1986) Bicyclo-prostaglandin E2 metabolite in congestive heart failure and relation to vasoconstrictor neurohumoral principles. Am J Cardiol 57:619–623
9. Leithe ME, Margorien RD, Hermiller JB, Unverferth DV, Leier CV (1984) Relationship between central hemodynamics and regional blood flow in normal subjects and in patients with congestive heart failure. Circulation 69:57–64
10. Domingo E, Cardus D, McTaggart W (1985) Simultaneous central and peripheral cardiovascular impedance recordings. Automedica 6:193–208
11. Domingo E, Cardus D, Mctaggart W (1987) Exercise cardiac output by the electrical impedance technique. Automedica 9:287–300
12. Packer M (1993) How should physicians view heart failure? The philosophical and physiological evolution of three conceptual models of the disease. Am J Cardiol 71:3C–11C
13. Townend J, Lote C, Littler W, Davies M (1993) Placebo controlled study of the effect of inhibition of prostaglandin synthesis with indomethacin on regional blood flow incongestive heart failure (abstract). JACC 21:467A
14. Malik KU, McGiff JC (1976) Cardiovascular actions of prostaglandins. In: Karim SMM (ed) Prostaglandins: physiological, pharmacological and pathological aspects. MTP Press, Lancaster, pp 103–200
15. Gilabert MR, Ballester R, Soler-Soler J, Domingo E (1993) Influencia de la aspirina en la distribucion regional del flujo sanguineo durante el ejercicio en la insuficiencia cardiaca (abstract). Rev Esp Cardiol 46 [Suppl I]:16
16. Gilabert MR, Domingo E, Soler-Soler J (1991) Aspirin and prostaglandin axis: effect on maximal exercise in healthy (abstract). Chest 100:S113
17. Staessen J, Cattaert A, Fagard R et al (1984) Hemodynamic and humoral effects of prostaglandin inhibition in exercising humans. J Appl Physiol 56 (1):39–45
18. Domingo E, Gilabert MR, Alio J, Soler-Soler J (1991) Effect of drugs on a noninvasive index of arterial compliance in healthy and heart failure patients. Cathet Cardiovasc Diagn 24:93–98

19. Sanders TM, Werner A, Bloor M (1976) Visceral blood flow distribution during exercise to exhaustion in conscious dogs. J Appl Physiol 40:927–931
20. Finn WF, Arendshorst WJ (1976) Effect of prostaglandin synthetase inhibitors on renal blood flow in the rat. Am J Physiol 231:1541–1545
21. Chow CC, Alemayehu A, Mangino MJ (1989) Prostanoid in regulation of postprandial yeyunal hyperemia and oxygen uptake. Am J Physiol 257:G798–G808
22. Leier CV (1992) Regional blood flow in human congestive heart failure. Am Heart J 124(3):726–738
23. Schwartz D, Kornowski R, Lehrman H et al (1992) Combined effect of captopril ans aspirin in renal hemodynamics in elderly patients with congestive heart failure. Cardiology 81(6):334–339
24. Ueyama M, Nishimura H, Kubota J, Kawamura K (1993) Hemodynamic and neurohumoral responses to exercise in patients with congestive heart failure. Jpn Circ J 57(5):411–417
25. Zelis R, Flaim SF (1982) Alterations in vasomotor tone in congestive heart failure. Prog Cardiovasc Dis 24:437

Resuscitation of Trauma Patients

A. Gullo and G. Berlot

Introduction

In Western countries trauma is the leading cause of mortality and disability among those aged under 30 years [1]. According to official statistics only in Italy do more than 8000 people die every year in road accidents, and many more remain severely disabled. While there are differences in the types of injuries between the United States and Europe (basically much more frequent open trauma, such as gunshot and stab wounds, in the former), this epidemic entails extremely high human, economic, and social costs. It is conceivable that a more comprehensive diagnostic and therapeutic approach could reduce this burden.

Mortality among trauma patients has a trimodal distribution [2]: (a) Roughly 50% die in the first hour after the initial event due to injuries incompatible with life. The adoption of preventive measures (speed limits, airbags, helmets, etc) appears the only effective way to reduce mortality in this very early phase. (b) Another 30% die within a few hours after the trauma due to factors that are at least partially preventable, including delayed triage and transportation to referral centers, missed or underevaluated lesions, etc. It is likely that the outcome of many of these could be improved by a more comprehensive resuscitative strategy as well by the improvement in management both outside of hospital and in hospital [3]. (c) The remaining deaths are due to delayed and indirect complications of the initial event, generally associated with the development of sepsis and multiple organ dysfunction syndrome. It is likely that in these patients too, an early and aggressive treatment could reduce mortality because the factors responsible for sepsis and its associated cardiorespiratory and metabolic derangements are probably present from the very early posttraumatic phase [4]. Thus the appropriate treatment ot trauma patients should ideally be target-oriented against both the more immediate life-threatening alterations and their delayed consequences.

Pathophysiology

Severe trauma is usually associated with the development of hypovolemia due to bleeding and/or tissue edema. Should a fall in oxygen transport (DO_2)

ensue, it could be futher worsened either by acute concomitant complications causing hypoxemia, such as head injuries, thoracic trauma, or lung aspiration or by preexisting cardiopulmonary disorders. With reduced DO_2 the acute increased peripheral oxygen requirements driven by the neurohormonal response to the trauma [4] can be initially matched by a parallel increase in tissue oxygen extraction (O_{2er}). However, when this mechanism is no longer able to support aerobic metabolism in the face of a continuously falling DO_2, tissue hypoxia occurs, associated with the reduction of oxygen utilization (VO_2).

There is a wide interorgan difference in O_{2er} capabilities, causing some regions to be more sensible to ischemia than others: in certain shock models the gut has been demonstrated to be more sensitive to hypoxia than muscle, due to the particular array of the mucosal microvascular network [5]. Several investigations have demonstrated that in critically ill septic and trauma patients the reduction in splanchnic perfusion is limited since the early postinjury phase, as reflected by a low gastric intramucosal pH (pHi), and that persistently low values are associated with a worse prognosis [6–9]. Especially in the very early stage, systemic hemodynamic variables fail to reflect splanchnic hypoperfusion [10]. In trauma patients the integrity of the gut mucosa is at risk not only during the immediate posttraumatic period. The intestinal epithelial cells appear particularly at risk of ischemia-reperfusion injury when oxygen is supplied again during the resuscitation phase, due to their high levels of the enzyme xanthine dehydrogenase contained in the mucosa.

Actually, the conversion of xanthine dehydrogenase to xanthine oxidase, which is the very initial step of the biochemical cascade leading to the production of oxygen free radicals, takes as few as 10 s in the gut, compared with 8 min in the cardiac muscle and 30 min in the kidney, spleen, and lung [11]. Thus, it has been hypothesized that the hypoxia-associated breakthrough of the intestinal epitelial lining, possibly further aggravated by the generation of oxygen free radicals during the reperfusion phase, can induce the escape of the intestinal content into the bloodstream, leading to the activation of many mediators, including tumor necrosis factor, interleukins 1–10, and platelet-activating factor, which can further enhance the initial damage [12]. However, although the translocation of bacteria or bacteria-derived substances has been demonstrated in various models of trauma [13, 14], its role humans has not yet been established [15]. In a number of clinical settings several authors have associated an absolute or relative reduction in the splanchnic blood flow with the presence of either bacteria or endotoxin in the bloodstream or in the lymphatics draining the gut. Rush et al. [16] observed a high rate of positive blood cultures in 56% of trauma patients in shock within 3 h after admission, in some cases associated with endotoxemia. Reed et al. [17] demonstrated that splanchnic nodes of a high percentage of patients undergoing emergency surgery after abdominal trauma contained considerable amounts of live enteric bacteria; however, this finding was not associated with an increased rate of complications, including sepsis and septic shock, nor with an increased length of stay in the hospital. A reduction in gastric intramucosal pH associated with measurable levels of endotoxin in the portal blood has also been demonstrated in patients undergoing heart surgery, probably as the result of reduction in blood flow during cardiopulmonary bypass surgery [18]. Assays of other mediators in the bloodstream of trauma patients have yielded conflicting results. High blood levels of tumor necrosis factor have been measured in the serum of patients with burns and trauma [19, 20]; however, other investigators failed to detect a rise in blood tumor necrosis factor levels in the immediate posttraumatic phase [21].

As noted above, mortality in the late phase is closely associated with infections and their complications. Bacteria and fungi can invade the organism of trauma patients through many different routes in addition to the gut, including open wounds, bruises, intravascular lines, urinary catheters, and tracheal tubes. Moreover, along with the increased exposure to germs, a number of immunological abnormalities have been demonstrated in patients with trauma and shock, including reduced processing of invading germs by T-cells and macrophages, dysfunction of B-cells and neutrophils, and production of middle and high molecular weight immunosuppressive substances [2]. The extent of the disturbance in cellular immunity is more related to the severity of trauma than to the duration of shock or to the resuscitation protocol, as recently demonstrated in a model of hemorrhagic shock [23]. Thus, it appears that the early-onset hypovolemia-associated reduction in DO_2 associated with the limited O_{2er} capabilities of the gut can jeopardize the integrity of the enteric mucosal epithelial lining, and this initial damage can be further aggravated by the oxygen free radicals generated during the resuscitation phase [24]. If the patient survives the initial insults, the ensuing intestinal hypoxic injury leading to the loss of mucosal barrier integrity, possibly associated with invasion by other germs from other routes of entry, could be the primer for the delayed-onset consequences of severe trauma.

Diagnostic Approach

The evaluation of critically ill trauma patients (i.e., those with Injury Severity Score of 15 or higher) [2] should include a thorough physical assessment, keeping in mind that in these patients the presence of multiple injuries is more a rule than an exception. The basic goals of the initial evaluation are particularly aimed at the assessing the neurological conditions and identifying possible hidden sources of bleeding. Neurological assessment is usually based on the Glasgow Coma Scale. Unfortunately, in trauma patients, overall neurological performances and thus the Glasgow Coma Scale score can be affected by other factors than the underlying brain injuries, including alcohol or drug abuse, hypotension, anemia, and hypoxemia. This limitation can be overcame by radiological investigations, which include computed tomography and magnetic resonance imaging of the brain and spine. It should be kept in mind that initial, relatively mild lesions can worsen in the following weeks or months [25]. In selected patients cerebral angiography can provide information not obtainable with computed tomography, such as carotideal dissection following blunt trauma [26].

In addition to the conventional classifications in intra- or extracerebral hematomas, a more in-depth assessment can be obtained by evaluating the severity of the concomitant diffuse brain lesions, as recently proposed by Marshall et al. [27]. The search for sources of bleeding can be a challenging and time-consuming task. In the absence of obvious hemorrhagic lesions, hidden hemorrhagic lesions in patients surviving until admission to the trauma center are usually in the pleural space, thigh, or abdomen, including the retroperitoneal space and and the pelvis [1]. Although universally accepted guidelines are desiderable, the diagnostic workup varies from one institutions to another, mainly due to the different technological and cultural backgrounds of the teams involved in the care of severely injured patients. The choice of the investigations should be based on the clinical conditions, trying to achieve the maximal benefit in terms of information obtained with the minimal cost in terms of time saved. In hemodynamically stable patients thoracic and abdominal computed tomography supplies valuable information, which can be supplemented by angiography. In unstable patients a peritoneal tap and lavage can be performed in minutes, thereby gaining precious time for life-saving surgical procedures. In addition to clinical conditions, the available technology also affects the priority of investigations. Compared with only few years ago, more recent and rapid computed tomography scans allow more information to be gathered in less time: in our institution contrast-enhanced scans of the head, thorax, and abdomen are feasible in less that 15 min on a 24-h basis: this has decreased the use of ultrasonography, which suffers from a number of limitations and from low interobserver diagnostic accuracy.

Treatment

The treatment of trauma patients should be divided into two different phases, even though large overlaps exist, and the border between the two is not well defined. This is mainly because, as noted above, the pathophysiological derangements leading to the delayed-onset complications are probably already established when the patients enter the emergency ward. The first phase starts as soon as the rescue team arrives on the scene, and the basic principles during this phase do not differ substantially from those established by the American Heart Association for the treatment of patients suffering from cardiac arrest [28]. These consist of the restoration and mainteinance of airways patency (A), the supply of oxygen and ensuring of breathing (B), and the administration of fluids and vasoactive agents to support circulation (C) – thus the "ABCs" of resuscitation. However, the latter point has recently been questioned by Brickell et al. [29], who observed in hypotensive patients with penetrating torso injuries a better outcome and fewer postoperative complications among those rushed directly to the emergency ward than among those who received volume-resuscitation during the prehospital phase. The authors hypothesized that the worse prognosis of the treated group could be attributed to the increased blood loss due to the higher arterial pressure associated to the volume resuscitation. It is not clear whether the same considerations apply to patients with blunt torso injury, which is much more common in Europe, who are currently treated with volume resuscitation via two or more large-bore peripheral intravenous lines, according to the guidelines established by the American College of Surgeons [30] through its Advanced Trauma Life Support course.

The choice of resuscitation fluids has been fiercely debated. The American College of Surgeons currently recommends that the initial fluids be isotonic crystalloid infusion, which are cheaper than colloids and can provide the electrolytes lost from the interstitial space. Moreover, these appear to have been as effective as colloids in the resuscitation of trauma patients, and their effect on the extravascular lung water has not been definitively proven, although colloid-treated patients are resuscitated more rapidly and require smaller amount of fluids [31]. The administration of whole blood or packed red cells is indicated in the presence of tissue hypoxia despite volume resuscitation, severe anemia, or hemodynamic instability. If a delay of 45 min is not of major concern, the transfusion of fully typed and cross-matched blood is indicated; otherwise type-specific but not cross-matched blood can be prepared and transfused in 10 min. In extreme situations, type O blood is indicated. In women of childbearing age type O, RH^- blood is indicated to avoid sensitization and possible future complications during childbirth, whereas in both men and women beyond the the childbearing age blood type O, RH^+ can be given. If available, autotransfuion can reduce the needs for autologous blood, thereby reducing the risk of transmissible diseases as well [32]. In the more advanced phase of in-hospital resuscitation, surgery plays a major role. The major priority is commonly assigned to interventions for intracranial mass lesions, which are followed by repair of abdominal, thoracic, and peripheral vascular injuries [1]. The immediate surgical approach of most orthopedic injuries is associated with a reduction in postoperative infection rate and morbidity, due to the earlier mobilization and better pulmonary toilette [33]. Maxillofacial injuries have the lowest priority, together with selected skeletal injuries, such as acetabular fractures, which require stabilization before final treatment.

Conclusions

The management of trauma patients requires such a large amount of theoretical background and practical expertise that a single specialist cannot be asked to handle it. Treatment during the very first hours after the occurrence of the trauma requires close cooperation between the emergency physicians, who are expected to perform the ABCs of resuscitation on the scene, and the anesthesiologist, who keeps the patients alive while other specialists, including radiologists and surgeons, carry out the diagnostic workup and perform life-saving procedures. Several hours later state-of-the-art treatment is required of the intensivist, which should include an in-depth knowledge of physiology, nutrition, infectious diseases, etc. It is clear that only a multidisciplinary approach can master this formidable task.

References

1. Trunkey DD (1983) Trauma. Sci Ann 282:298–235
2. Trunkey DD (1991) Initial treatment of patients with severe trauma. N Engl J Med 324:1259–1263
3. Stocchetti N, Pagliarini G, Gennari M et al (1994) Trauma care in Italy: evidence of in-hospital preventable deaths. J Trauma 36:401–405
4. Weissman C (1990) The metabolic response to stress: an overview and update. Anesthesiology 73:308–327
5. Fink MP (1991) Gastrointestinal mucosal injury in experimental models of shock, trauma and sepsis. Crit Care Med 19:627–641
6. Ruomen RMH, Vreugde JPC, Goris JA (1994) Gastric tonometry in multiple trauma patients. J Trauma 36:313–316
7. Chang MC, Cheatham ML, Nelson LD, Rutherford EJ, Morris JA (1994) Gastric tonometry supplements information provided by systemic indicators of oxygen transport. J Trauma 37(3):488–494
8. Doglio GR, Pusajo JF, Egurrola MA et al (1991) Gastric mucosal pH as a prognostic index of mortality in critically ill patients. Crit Care Med 19:1037–1040
9. Gutierrez G, Palizas F, Doglio G et al (1992) Gastric intramucosal pH as a therapeutic index of tissue oxygenation in critically ill patients. Lancet 339:195–199
10. Maynard N, Bihari D, Beale R et al (1993) Assessment of splanchnic oxygenation by gastric tonometry in patients with acute circulatory failure. JAMA 270:1203–1210
11. Grace PA (1994) Ischemia-reperfusion injury. Br J Surg 81:637–647
12. Deitch EA (1992) Multiple organ failure: pathophysiology and potential future therapy. Ann Surg 216:117–134
13. Morris SE, Navaratnam T, Townsend CM et al (1989) Decreased mesenteric blood flow independently promotes bacterial translocation in chronically instrumented sheep. Surg Forum 40:88–92
14. Jones WG II, Minei JP, Barber AE et al (1990) Bacterial translocation and intestinal atrophy after thermal injury and burn wound sepsis. Ann Surg 211:399–403
15. Edmiston CE, Condon RE (1991) Bacterial translocation. Surg Gynecol Obstet 173:73–83
16. Rush BF, Sori AJ, Murphy TF, Smith S, Flanagan JJ, Machiedo GW (1987) Endotoxemia and bacteremia during hemorrhagic shock. The link between trauma and sepsis? Ann Surg 207:549–554
17. Reed L, Martin M, Mangano R, Newson B, Kocka F, Barrettt J (1994) Bacterial translocation following abdominal trauma in humans. Circ Shock 42:1–6
18. Andersen LW, Landow L, Baek L, Jansen E, Baker S (1993) Association between gastric imtramucosal pH and splanchnic endotoxin antibody to endotoxin and tumour necrosis factor concentration in patients undergoing cardiopulmonary bypass. Crit Care Med 21:210–217
19. Takayama TK, Miller C, Szabo G (1990) Elevated tumour necrosis factor production concomitant to elevated prostaglandin E2 production by trauma patient's monocytes. Arch Surg 125:29–35
20. Marano MA, Fong Y, Moldawer LL et al (1990) Serum cachectin/tumour necrosis factor in critically ill patients with burns correlates with infection and mortality. Surg Gynecol Obstet 170:32–38

21. Hoch RC, Rodriguez R, Manning T et al (1993) Effects of accidental trauma on cytokine and endotoxin production. Crit Care Med 21:869–845
22. Abraham E (1989) Host defense abnormalities after hemorrhage, trauma and burns. Crit Care Med 17:934–939
23. Schmand JF, Ayala A, Chaudry IH (1994) Effects of trauma, duration of hypotension, and resuscitation regimen on cellular immunity after hemorrhagic shock. Crit Care Med $_2$:1076–1083
24. McCord JM (1985) Oxygen-derived free radicals in postischemic tissue injury. N Engl J Med 312:159–163
25. Berlot G, Viviani M, Gullo A (1992) Traumatic carotideal dissection after blunt cervical injury: an elusive clinical entity. Am J Emerg Med 10(4):396–398
26. Berlot G, Viviani M, Magnaldi S, Gullo A (1995) Delayed traumatic cervical cord transection: case report. Am J Emerg Med (in press)
27. Marshall LF, Bowers Marshall S, Klauber MR et al: A new classification of head injury based on computed tomography. J Neurosurg 119(75):S14–S20
28. Anonymous (1992) Standards and Guidelines for Cardiopulmonary Resuscitation (CPR) and Emergency Cardiac Care (ECC) JAMA 268 (16):2172–2298
29. Brickell H, Wall MJ, Pepe PE et al (1994) Immediate versus delayed fluid resuscitation for hypotensive patients with penetrating torso trauma. N Engl J Med 331:1105–1109
30. American College of Surgeons, Committee on Trauma (1993) Advanced trauma life support program for physicians: instructor manual. American College of Surgeons, Chicago, pp 75–110
31. Falk JL, O'Brien JF, Kerr R (1992) Fluid resuscitation in traumatic hemorrhagic shock. Crit Care Clin 8:323–340
32. Dawson RB (1987) Transfusion: volume expansion, oxygen transport, hemostasis, transufion reactions, autotransfusion. In: Cowley RA, Conn A, Dunham CM (eds) Trauma care: medical management. Lippincott, New York, p 131
33. Seibel R, LaDuca J, Hassett JM et al (1985) Blunt multiple trauma (ISS), femur traction and the pulmonary failure syndrome. Ann Surg 202:283–295

Subject Index

Springer-Verlag and the Environment

We at Springer-Verlag firmly believe that an international science publisher has a special obligation to the environment, and our corporate policies consistently reflect this conviction.

We also expect our business partners – paper mills, printers, packaging manufacturers, etc. – to commit themselves to using environmentally friendly materials and production processes.

The paper in this book is made from low- or no-chlorine pulp and is acid free, in conformance with international standards for paper permanency.

SPRINGER NATURE

GPSR Compliance

The European Union's (EU) General Product Safety Regulation (GPSR) is a set of rules that requires consumer products to be safe and our obligations to ensure this.

If you have any concerns about our products, you can contact us on ProductSafety@springernature.com

In case Publisher is established outside the EU, the EU authorized representative is:

Springer Nature Customer Service Center GmbH
Europaplatz 3
69115 Heidelberg, Germany

FSC
www.fsc.org
MIX
Papier | Fördert gute Waldnutzung
FSC® C083411

Kolibri 360 GmbH
Ferdinand-Jühlke-Straße 7
99095 Erfurt, Deutschland
produktsicherheit@kolibri360.de